Biology of Genus Boswellia

Ahmed Al-Harrasi • Abdul Latif Khan
Sajjad Asaf • Ahmed Al-Rawahi

Biology of Genus Boswellia

Springer

Ahmed Al-Harrasi
University of Nizwa
Nizwa, Oman

Abdul Latif Khan
University of Nizwa
Nizwa, Oman

Sajjad Asaf
University of Nizwa
Nizwa, Oman

Ahmed Al-Rawahi
University of Nizwa
Nizwa, Oman

ISBN 978-3-030-16724-0 ISBN 978-3-030-16725-7 (eBook)
https://doi.org/10.1007/978-3-030-16725-7

This Springer imprint is published by the registered company Springer Nature Switzerland AG
The registered company address is: Gewerbestrasse 11, 6330 Cham, Switzerland

Foreword

Frankincense is precious. It is a resin that is used worldwide in cultural and religious ceremonies, not only in the past but still today. It has been traded for the last 5000 years and has supported happiness and prosperity in many cultures and regions.

Who has never heard of the 'Frankincense Trail', long distance camel transport routes from southern Arabian Peninsula, and possibly the Horn of Africa, towards the large cities in the Middle East, Greek and Roman Classical Empires and the Indian civilizations? Frankincense was a main product transported along these trails and traded for other precious goods. Who has never heard of 'gold, frankincense and myrrh', the famous precious gifts that were brought by the three kings to the newborn Christ just over two thousand years ago? Who has never heard of frankincense being an essential part of many worldwide used perfumes and an essential part of traditional medicines both in their source countries and in faraway countries like China?

Frankincense is collected from trees and shrubs in the dry tropics of the Arabian Peninsula, the sub-Sahel region in Africa, the Horn of Africa and the Indian subcontinent. The genus thus has a very wide distribution, and of the 19 species, there are five that are produced for local, regional, national and international markets. The species taxonomy is still debated, the distribution of *Boswellia* trees and their abundances are known for some geographical locations but not for others, the resin chemistry is now more understood but not the genetics. So far, almost all of the resin is being collected from the wild populations in far away and mostly environmentally harsh locations. These populations are under increasing threats due to unkind cut and maltreatment by unprofessional harvest of the resin. In addition, the young trees are being eaten by cattle, goats, sheep and camels as well as are being burnt by frequent fires. This needs urgent attention.

My personal involvement as a biologist with frankincense started over 20 years ago when an Eritrean student asked me to be his PhD advisor. Since then I have been working much in Eritrea and Ethiopia on the biology and sustainable management of frankincense plants and the vegetation in which they thrive. As so little was known, all results were new and exciting, and I kept on working and getting more and more involved in frankincense. For years I looked forward to the possibility to visit Oman, the world-famous land of frankincense, where *Boswellia sacra* grows. In late 2018 the great opportunity emerged as some great Omani frankincense researchers organized the first international frankincense scientific symposium in Muscat. And the book you are reading now is written by these great researchers.

I visited Dhofar, the land of frankincense, and travelled to several frankincense-growing locations, some along the coast and others in the interior and on the mountains. The *Boswellia sacra* trees are impressively green and healthy with fascinating white-pinkish flowers. The landscape is open, dry, rocky and harsh for sure. Several large-sized trees had toppled over, probably the result of strong winds during the past rainy season. Small plants were growing here and there; they did not abound but maybe enough to replace the older ones. I was excited to see all this and discuss it with my fellow travellers, both frankincense researchers as I am and frankincense collectors and producers. Lots of exciting questions remain to be addressed, but we all agreed that we need to make sure that this species will be treated with respect, that we need to help to make long-term sustainability possible.

This new book is the first scientific book on biology of genus *Boswellia*, with its in depth information focusses on many undiscovered aspects of *Boswellia sacra* and its resin. It highlights the genetics and genomics, and much of the details that the Omani group is concentrating on, but not only that. I think it is very important and informative that these specialty topics are embedded in a broad framework of species descriptions (look at the great pictures in this book) and species ecology. Much emphasis is put to the collection of the resin: how the plants are being tapped, what are the biological consequences thereof and how much resin is collected. Production and trade, not only nowadays, but also in the past, is brought to the front.

For me, a person with a very broad interest in all aspects of frankincense, this is really exciting and rewarding. And I am convinced that many people are interested in several of the topics tackled in this book. I commend the authors with such a detailed book reflecting their exciting and new work on this famous species and its highly valued resin frankincense.

Wageningen University & Research,
Wageningen, The Netherlands
January 2019

Frans Bongers

Introduction

Frankincense has long been associated with different cultures, civilizations and religions. In the Arabian culture, particularly in Oman and Yemen, frankincense is popular in social and religious celebrations, while in western cultures, its characteristic smell is associated with the Catholic church. Frankincense is burned on hot charcoal, and incense smoke is emitted. In Christianity, frankincense is mentioned 22 times in the Bible (Duke, 2008). The ancient Roman, Egyptian and Greek civilizations knew about the therapeutic importance of frankincense, and very early in human history, frankincense trading routes were established starting from the southern part of the Arabian Peninsula, primarily from Dhofar in southern Oman and Hadramaut (Yemen), to the capitals of the western world (Al-Ghassany, 2008).

The word 'frankincense' is derived from the old French word franc-encens, meaning 'pure incense' (Skeet, 1963), or, literally, 'free lighting' (Walker, 1957), and it has appeared in different civilizations under different names. The Arabic word for frankincense is *Luban*, which is derived from the Semitic root denoting whiteness and implying purity. The Hebrew name is *Lebona*, the Greek is *Libanos* or *libanotos* and the Latin is *tus* (Groom, 1981; Walker, 1957). The ancient Egyptian name was *neter-sent* (Tucker, 1986). In India, it is called *salai gugga*; in Socotra, which is one of the centres of endisms of *Boswellia*, which hosts seven rare species, it is called *am eiro*. In Somalia, frankincense from *Boswellia carterii* is known as *moxor*, while frankincense from *Boswellia frereana* is known as *jagcaar*, and frankincense from *Boswellia papyrifera* is known as *boido*.

The famous frankincense trade was well documented because of its importance in linking different civilizations during ancient times. The earliest reports concerning the frankincense trade route go back to the old Babylonians, when the caravan roads crossed between India, Arabia and Syria. In the Babylonian settlement of Sippar (ca. 2250 BC), the merchants exchanged goods for natural products, including frankincense and myrrh. Both frankincense and myrrh resins were used by old Egyptians for fumigation, as revealed by an analysis of archaeological samples. Between the third century BCE and the second century CE, the incense trade flourished from South Arabia to the Mediterranean through Incense Road. Frankincense and myrrh were carried on camels from southern Oman (Dhofar), Yemen and Hejaz to the Mediterranean ports of Gaza.

The exceptional importance of frankincense explains the interest of ancient civilizations in this material. Among several other uses, its pharmacological use rises to the top. Frankincense was connected to the Indian philosophy of Ayurveda, and it comes from *Boswellia serrata* and was used to treat a variety of diseases (Ammon, 2006). Furthermore, the scripts of the Ebers Papyrus mention frankincense as a drug.

There have been 19 reported species in the *Boswellia* genus since the discovery of the first species by the Scottish botanist John Boswell in 1807 until the last discovery of *Boswellia bullata* in 2001 on Socotra island. There has been much controversy in the literature with regard to the number of species and confusion about the original species and synonyms. For example, *Boswellia carterii* has long been confused with *Boswellia papyrifera*, and *Boswellia carterii* was assumed to be different from *Boswellia sacra*. This confusion is likely due to the use of commercial samples that were purchased from local markets without proper taxonomic identification. In this book, we adopt the species listed by Eslamieh in his book *The Genus Boswellia*. *Boswellia carterii* is a synonym of *Boswellia sacra*, but *Boswellia microphylla* is a different species from *Boswellia neglecta*, and *Boswellia popoviana* is different from *Boswellia nana* (Eslamieh, 2010). In this book, we have also omitted *Boswellia madagascariensis* from the list of *Boswellia* species because it is now considered independent.

Unfortunately, a similar observation was noted with regard to the chemical components of resins from different species, in particular the volatile constituents. Misleading results have been encountered in the past literature and have been quoted and requoted, leading to confusion.

The natural habitat of *Boswellia* trees (trees and shrubs) varies between dry regions as represented by Sudan, Ethiopia, Somalia, Yemen and Oman and humid habitats as represented by India and Sri Lanka. In southern Oman (Dhofar), the resin produced by the *Boswellia sacra* tree varies depending on the geographical location (mountainous and coastal areas) of the tree. This variation is reflected in the commercial grades of the frankincense produced from this tree.

Frankincense is obtained from frankincense trees through careful incisions into the trunk of the tree in which a small strip of the bark is peeled off (cutting should be performed through the cambium) to allow the milk-like substance to ooze out from the trunk and be solidified by exposure to air (Fig. 1). These exudates are likely to be a result of a defensive mechanism through which the tree heals wounds from insect attacks and possibly helps to reduce water loss. The best quality frankincense is harvested after the second incision.

Unkind cuts (wide and deep) and over-tapping (excessive incisions) will render the tree vulnerable to insects and will eventually kill the tree. Interestingly, for a few species, including the coastal *Boswellia sacra* tree, and on certain occasions, the resin oozes out naturally due to the expansion of the trunk and the cracking of the cambium, which will ultimately lead to a rupture in the resin canals. Despite the geographical locations and the habitat of the tree, the method for collecting the resin is the same in all frankincense-producing countries. There are 19 species of

Fig. 1 Stages of harvesting frankincense and different grades of resin obtained from *Boswellia sacra*

frankincense that are geographically distributed throughout South Arabia, Africa and India, as will be discussed in Chap. 2.

The chemical profile of the resin varies significantly from one species to another. While some species are rich in monoterpenes, others are rich in diterpenes and triterpenes. The resin generally possesses fascinating structural diversity. In the essential oil of *Boswellia* from all the species, there have been more than 300 components identified to date (Mertens, Buettner, & Kirchhoff, 2009), and the resin contains more than 100 diterpenes and triterpenes (Al-Harrasi et al., 2018). The botanical origin, geographical location, time of harvest and other environmental factors contribute collectively to the overall composition and hence the biological activity of the resin.

With recent advances in modern medicine, the pharmacological importance of frankincense as a drug has declined, whereas its social and religious values have been maintained. This trend was observed by the disappearance of the term 'Olibanum' from pharmacopoeias in the middle of the twentieth century. However, with the development of preclinical and clinical studies at the end of the last century, frankincense has regained its value when some studies supported its potency in treating a variety of diseases, including rheumatoid arthritis, ulcerative colitis, bronchial asthma and multiple sclerosis (Al-Harrasi et al., 2018).

The frankincense extracts or pure compounds isolated from various *Boswellia* species, in particular triterpenes and diterpenes, have demonstrated superior anti-inflammatory and anticancer activities (Al-Harrasi et al., 2018). The majority of the reported activity of frankincense is due to its rich content of boswellic acids. Another fascinating class of cembranoids, primarily incensole and incensole acetates, displays exceptional anti-inflammatory and anti-depression activities due to their ability to activate ion channels in the brain to alleviate anxiety or depression. In parallel to the advances in clinical studies, several pharmaceutical and cosmetic frankincense-based products have appeared on the market. With the development of olfactometry science, a remarkable achievement has been made with regard to the identification of the fragrant components of frankincense, which has in turn facilitated the development of frankincense-based fragrances.

This is the first book on the biology of the frankincense tree. It is noteworthy that the chemistry and bioactivity of boswellic acids and other terpenoids from the *Boswellia* genus have been covered in our recently published book (Al-Harrasi et al., 2018), and the applications of frankincense in western medicine have been discussed in detail by Hermann P. T. Ammon in his book *Weihrauch-Anwendung in der westlichen Medizin*' (Ammon, 2006). Furthermore, the horticulture of the genus *Boswellia* was well described in a book by Eslamieh (Eslamieh, 2017).

Chapter 1 describes the human uses of frankincense during ancient times and in different civilizations. This chapter also highlights the frankincense trading route and its importance in cultural, civilizational and economic activities across different continents. Chapter 2 emphasizes the *Boswellia* species, their taxonomic identifications and details about the *Burseraceae* family. Chapter 3 reports the production of the resin inside the tree and its transportation to different parts of the stem. Chapter 4 explains the physiological tapping or wounding of the *Boswellia* tree and its ensuing physiochemical-molecular responses via phytohormones, essential biochemicals and metabolites. Chapter 5 illustrates the recent trends in the ex situ conservation of *Boswellia* through tissue culture and bud propagation methods. Chapter 6 describes the recently reported trends in genetic diversity assessments of various populations of *Boswellia sacra* and *Boswellia papyrifera*. It also explains the conservation threats confronted by these populations using various molecular markers and detailed phylogenetic differentiation. Chapter 7 presents the genomics of *Boswellia sacra* and its gene map. This chapter provides detailed information on designing molecular markers for understanding and assessing population gene flow and diversity. Chapter 8 describes the bacterial and fungal rhizosphere communities living with both cultivated and wild *Boswellia sacra* tree populations using next-generation sequencing approaches. Chapter 9 explains the diversity and abundance of endophytes (bacteria or fungi), which provide a diverse hub of bioactive secondary metabolites, phytohormones, extracellular enzymes and essential nutrients. This information not only assists in understanding the role of associated microorganisms but also helps in understanding the tree life and evolution. Chapter 10 summarizes the chemical composition of resins harvested from the 19 species in the *Boswellia* genus, which will be attractive to biologists, chemists, pharmacologists and medicinal chemists due to their fascinating structural diversity.

Nizwa, Oman

Ahmed Al-Harrasi
Abdul Latif Khan
Sajjad Asaf
Ahmed Al-Rawahi

References

Al-Ghassany, D. (2008). the Land of Frankincense.
Al-Harrasi, A., Rehman, N. U., Khan, A. L., Al-Broumi, M., Al-Amri, I., Hussain, J., … Csuk, R. (2018). Chemical, molecular and structural studies of Boswellia species: β-Boswellic Aldehyde and 3-epi-11β-Dihydroxy BA as precursors in biosynthesis of boswellic acids. *PLoS One, 13*(6), e0198666.
Ammon, H. (2006). Boswellic acids in chronic inflammatory diseases. *Planta medica, 72*(12), 1100–1116.
Duke, J. A. (2008). *Duke's handbook of medicinal plants of Latin America*: crc press.
Eslamieh, J. (2010). Creating "Perfect" Boswellia. *Cactus and Succulent Journal, 82*(3), 126–131.
Groom, N. (1981). Frankincense and myrrh. A study of the Arabian incense trade. *Longman: London & New York, 285*, 96–120.
Mertens, M., Buettner, A., & Kirchhoff, E. (2009). The volatile constituents of frankincense–a review. *Flavour and Fragrance Journal, 24*(6), 279–300.
Skeet, W. W. (1963). An etymological dictionary of the English Language: London: Oxford University Press.
Tucker, A. O. (1986). Frankincense and myrrh. *Economic Botany, 40*(4), 425–433.
Walker, W. (1957). All the Plants of the Bible.

Contents

About the Authors

Ahmed Al-Harrasi received his BSc in Chemistry from Sultan Qaboos University (Oman) in 1997. Then, he moved to the Free University of Berlin from which he obtained his MSc in Chemistry in 2002 and then his PhD in Organic Chemistry in 2005 as a DAAD fellow under the supervision of Prof. Hans-Ulrich Reissig. His PhD work was on New Transformations of enantiopure 1,2-oxazines. Then, he obtained the Fulbright Award in 2008 for his postdoctoral research in chemistry for which he joined Prof. Tadhg Begley's group at Cornell University where he worked on the synthesis of isotopically labelled thiamine pyrophosphate. After a postdoctoral research stay at Cornell University in 2009, he started his independent research at the University of Nizwa, Oman, where he founded the Chair of Oman's Medicinal Plants and Marine Natural Products merging chemistry and biology research. He is currently a professor of organic chemistry and the vice chancellor for Graduate Studies, Research and External Relations at the University of Nizwa. He is also the founder and chairperson of the Chair of Oman's Medicinal Plants and Marine Natural Products. The budget of his interdisciplinary-funded projects exceeds seven million USD. He was a chair and invited speaker in many international conferences and is a referee for more than 15 international chemistry and biotechnology journals. He has authored and co-authored over 300 scientific papers, 1 book and 6 book chapters and taught many chemistry courses both at MSc and BSc levels.

Abdul Latif Khan received his BSc and MSc degree in distinction from the University of Peshawar, Pakistan. Later, he joined the Department of Chemistry, Kohat University of Science and Technology, Kohat, Pakistan, to complete his MPhil in Phytochemistry and JASSO Research Fellowship by the Gene Research Center University of Tsukuba, Japan. He was selected as honorary scholar by Kyungpook National University, South Korea, for his PhD degree in Plant Physiology. After his PhD, he remained as a postdoctorate fellow at the School of Applied Biosciences. He also worked as research professor at the Institute of Agricultural Science and Technology, Kyungpook National University, South Korea. He has published more than 150 research articles as a principal author and co-author in journals of international repute and of impact factor. His research focuses on plant molecular physiology during the activities of environmental and microbial antagonists and plant genome sequencing. His current projects include draft and chloroplast genomes of date palm, *Boswellia sacra*, pomegranate and various endemic medicinal plants of Oman. He is currently working as associate professor at the Natural and Medical Sciences Research Center, University of Nizwa, Nizwa, Oman.

Sajjad Asaf completed his Master in Botany (MPhil) from Kohat University of Science and Technology (KUST), Kohat, Pakistan, in 2012. He received a KNU Honour Scholarship for PhD studies in South Korea and completed his PhD (2017) from the Laboratory of Crop Physiology at Kyungpook National University, South Korea. His PhD research work was based on the physiology and genomics of plant growth-promoting endophytic bacteria and its role in environmental stress tolerance. Little is known about bacterial endophytes from arid land plants. Beyond his PhD research project, he is deeply committed to genome sequencing and the analysis of plant organelles and their comparison with nuclear genomes. He has been fortunate to obtain a wide range of experience in the field of chloroplast, mitochondria and microbial genomics. Moreover, he pursued postdoctoral research for 1 year in plant physiology and genomics at Kyungpook National University, South Korea. Currently, he is working as assistant research professor at the University of Nizwa, Oman. He is currently involved on the whole plant genome sequencing analysis, as he has experience in plant, chloroplast, mitochondrial and microbe's genome sequencing and data analysis. Therefore, comparative genomic analysis is the main focus of his current interest. Furthermore, he has the potential to analyse the data through different bioinformatics tools responsible for NGS data processing.

Ahmed Al-Rawhi earned his MSc and PhD in Plant Pathology from the University of California at Berkeley in 1992 and 1995, respectively. Earlier (1988), he had completed with Honours his BSc in Biological Science at North Carolina State University (Raleigh, USA). Upon his return to Oman, he became a lecturer at Sultan Qaboos University where he taught courses in the field of plant pathology and microbiology. He also conducted research in soil-borne pathogens, biocontrol and disease management programs and published articles in high-impact international journals. From December 1997 until May 2001, he had the honour of serving as minister of Agriculture and Fisheries and played, as such, a major role in developing various strategies and development plans for these vital sectors of the Omani economy. He served as member of the State Council from 2001 to 2011 and led important studies and reports for the government of Oman. He was also chairman of the Academic Foundation Committee for the University of Nizwa Project from 2000 to 2004, the year he was appointed as chancellor of the newly founded university. Throughout his tenure as chancellor, he embarked on bringing the project to reality and creating a functioning academic institution that adheres to institutionalization, quality standards, ethics and procedures. Under his dynamic leadership, thousands of Omani students graduated with bachelor's and higher degrees, and the university contributed strongly to community services and produced innovative research relevant to the needs of Oman. Ultimately, he was promoted to the academic position of founding professor by a scholarly independent Academic Committee in December 2006. Professor Ahmed Al Rawahi is a member of various national academic committees including the Education Council and Research Council. He was also a member of the American Phytopathological Society. In November 2000, he was greatly honoured to receive from HM Sultan Qaboos bin Said Oman Medal of Merits (2nd degree, civil). In December 2013, the University of Wisconsin Oshkosh awarded him the Honorary Doctorate Degree in recognition of his role as "a distinguished and visionary academic leader with an international voice and perspective".

Chapter 1
Frankincense and Human Civilization: A Historical Review

Etymology of Frankincense

In Arabic, the *Boswellia* tree and the resin it produces are called '*luban*' or '*loban*', which literarily describes the clear white exudate resin that is secreted by tapped trees and equates it with the '*laben*', or the 'milk' of a tree, hence, the origin of the 'libanum' and 'olibanum' used in ancient trade. '*Luban*' or '*loban*' for the plural and '*lebanah*' for a single tear droplet are not far from the Hebrew name '*levonah*' or '*lebonah*', which also indicate '*lavan*' or 'white' in Hebrew (Ben-Yehoshua, Borowitz, & Ondrej Hanuš, 2012; Coder, 2011). One of the important uses of frankincense since ancient times is as burning incense; thus, its English name 'frankincense' reflects these properties. The origin of this name comes from the Old French, with '*franc*' meaning pure or noble and '*encens*' meaning incense. Two types of Somalian frankincense, namely, '*Maidi*' and '*Bayo*', are the best.

Oleoresin (Tears of the Sun)

Frankincense gum resin is harvested by incising and scraping a piece of the periderm with a tool called a '*menguf*', which is an oval-shaped, scalpel-like instrument. The exposed surface is tapped to induce wounding, allowing the resin to ooze out. As it oozes out to the surface, the exudate appears to be similar to white milk in colour but is viscous as honey. Upon exposure to the air, it hardens, and its colour is transformed into a translucent to light amber, and the resin solidifies as pear-like tears or droplets known as 'tears of the sun'. After 2–3 weeks, those clumps of droplets are gently scraped off the tapped area and allowed to harden more for a few weeks before being marketed. During these collections, the tapped areas of the stem are re-tapped to induce further secretions from the tree. The resin is harvested during the dry season from October to December and from

A. Al-Harrasi et al., *Biology of Genus Boswellia*,
https://doi.org/10.1007/978-3-030-16725-7_1

March to May to avoid rain and foggy monsoon weather that affect the quality of the product. Mature trees are tapped for resin production when they reach an age of 7–10 years, and they continue to produce resin for 5–7 years before being allowed to rest (Coder, 2011; Hepper, 1969).

The quality of frankincense is dependent on its aroma, colour, purity and age. In Oman and Yemen, it is now produced just as it was in ancient times, only from wild, noncultivated trees. Notably, it is produced in other areas of Africa and India, and there was a reported project to cultivate the tree in Somalia for frankincense production prior to the civil war, with no available reports about how much resin was produced from the 'farmed' trees (Hepper, 1969; Thulin & Warfa, 1987; Tolera, Sass-Klaassen, Eshete, Bongers, & Sterck, 2013). In Dhofar, four types of frankincense have been recognized (35). A field study on the wild frankincense trees in Dhofar confirms that they are all members of *B. sacra* and they cannot be genetically differentiated into different cultivars (Coppi, Cecchi, Selvi, & Raffaelli, 2010; Raffaelli, Mosti, & Tardelli, 2003). Thus, this resin-gum variation is dependent on the tree locality as noted earlier by several field botanists, and the harvest time during a given season could contribute to whether the resin is of low or high grade within each type (Giuliani, Bini, & Lippi, 2012). There are three zones in Dhofar where wild frankincense trees are growing. One of these zones is the coastal plain from Ras Nus in the east to Ras Sajir and up to Rakhyut in the west, which extends approximately 8 km inwards from the shore. The trees tend to cluster in certain points along this plain. The frankincense produced in this zone is named '*shaabi*'. The trees in this growing zone are affected by high humidity during the year and dense fog and misty weather during the monsoon season; hence, the resulting frankincense is considered to be of a lower quality. The second zone is located to the south of the Samhan and Qara Mountains at their foothills and parallel to the shore. Frankincense trees are clustered in certain habitats such as north of Marbat and northeast of Risut. In fact, the ancient town known as '*Shisr*', which used to be a focal point of frankincense trade and the 'Incense Route or Road', is located within this zone, and consequently, the frankincense produced in this zone is named '*shasri*', after the town. The quality of this type is considered moderate. A third zone is located in the arid limestone hills north of the Qara Mountains called the '*Najid*'. The frankincense produced there is named '*nagidi*' and is considered the best type (Hepper, 1969). The trees that are farthest inland in this zone are known to produce the best grade of frankincense, which is distinguished by a special name '*hojari*', for which the highest price has long been offered. This name could be a reflection of the 'rocky stones' habitat (= *hajari* in Arabic) where the productive trees are located, or the name may be a recognition of how the precious resin tears are valued as gemstones. It is worth noting that the top quality frankincense type is produced by trees located away from the direct effect of the monsoon but that still receive some of its air-cooling effects.

Frankincense Trade Route

The characteristic smell and medicinal importance of incense gave rise to flourishing trade between South Arabia and the rest of the world. One of these routes involved Alexandria; from there, incense was processed and distributed to Rome and to the entire Mediterranean region during the Hellenistic-Roman period (Fig. 1.1) (Grohmann, 1933; Van Beek, 1960). The Hellenistic period covers the period of Mediterranean history between the death of Alexander the Great in 323 BC and the emergence of the Roman Empire as signified by the Battle of Actium in 31 BC and the subsequent conquest of Ptolemaic Egypt the following year. Some sources also revealed that incense was processed in Mesopotamia and India. Some of the distribution routes for these regions during the last centuries BC and the early

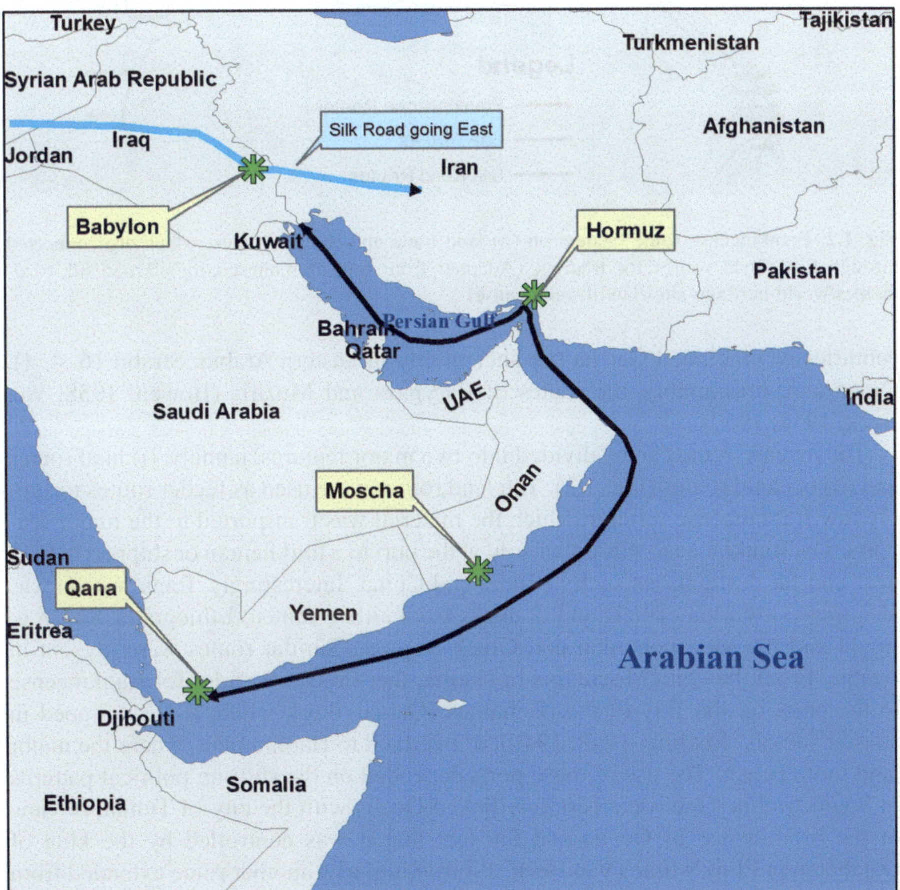

Fig. 1.1 Frankincense route from Qana to Hormuz and then Babylon. Adapted from the Middle Eastern Institute (https://www.mei.edu/sqcc/frankincense)

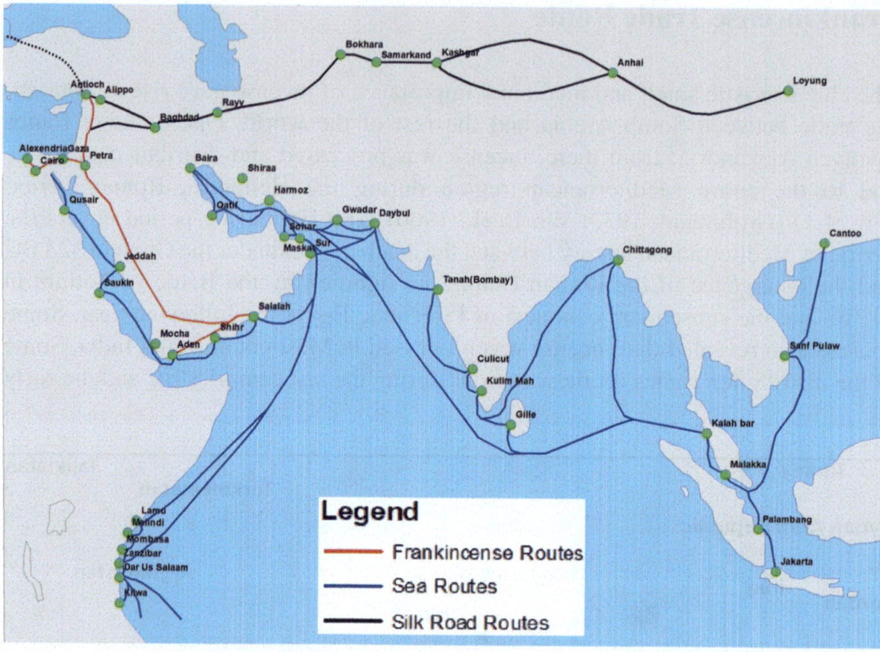

Fig. 1.2 Frankincense route feeders on (**a**) land trade and (**b**) sea routes, which are connected through Silk Road routes for trading. (Adapted from https://en.unesco.org/silkroad/silk-road-themes/world-heritage-sites/land-frankincense)

centuries AD included Gerrha (an ancient city of eastern Arabia; Strabo 16. 4. 4), Oman and, presumably, the routes of Barygaza and Muziris (Bowen, 1958; Van Beek, 1958, 1960).

The incense route can be divided into two major features, namely, (i) land routes and (ii) oceanic routes (Fig. 1.2). The land routes were used as feeder routes to supply raw frankincense through which the material was transported to the major seaports. The frankincense was either sold at the port to a middleman or shipped to both near and far destinations such as India and China. Interestingly, frankincense was also transported via camels and donkeys from either Yemen, Ethiopia or Sudan to Egypt and the Mesopotamian and Greek empires. Similar routes were present in Arabia. From the Qara Mountains in Dhofar, there was a corridor for frankincense to the ports on the Bay of the Saehalites [Qamar Bay], where it was shipped in coastal vessels (Ebeling, 1948, 1949), or overland to Hadhramaut, where the major land route began. The use of these ports depended on the shifting political patterns in South Arabia. One route certainly linked Ocelis with the city of Timna, in view of the prominence of Ocelis and the fact that it was controlled by the king of Qatabanin in Pliny's time (Van Beek, 1960). Similarly, another route extended from Beihan to Aden.

The probable existence of a cut-off route from Hadhramaut to Nejran crossing the Ramlet Sabatein and bypassing Qataban and Saba has been suggested by Bowen

29 following the work of Philby and Freya Stark. One route went to western Palestine, and another passed through northern Trans-Jordan and Teima to Mesopotamia. Throughout the entire history of the frankincense trade, the existence of a route connecting Gerrha and Hadhramaut is not at all surprising. That frankincense and myrrh were transported by sea in early times is known from Egyptian records, which describe trade with Punt [Somaliland]. From at least the Eleventh Dynasty, incense moved to Egypt over one basic route, from Somaliland to Quseir to Coptos and northwards on the Nile (Coppen, 1995). This route seems to have been used until the middle of the first millennium BC, when ports farther north replaced Quseir. Pliny lists alternate branches at the north end with (i) Berenice to Coptos and (ii) ltrsinoe to Alexandria being one of three routes (Fig. 1.2). To the Egyptian evidence for sea trade with Somaliland and almost certainly with South Arabia, we must also add the biblical account of Solomon's Phoenician-built merchant fleet, which operated out of Ezion-Geber (Glueck, 1940).

Frankincense Trade and Its Economic Importance

Historically, peak exports of over 3000 tons were shipped annually to the Roman Empire from southern Arabia during the first century CE (Coder, 2011). Most of those exports were assumed to be products of southern Arabia, and some might have been gathered and relocated from the Horn of Africa (Groom, 1981). During that era, the cost of all the merchandise imported by the Romans was estimated at 100 million sesterces, and 'more than half went on incense imported from Arabia' (Ben-Yehoshua et al., 2012). However, the benefits of the incense trade from transportation costs and from imposed taxing were shared by many nations along the Incense Road, from southern Arabia to Mesopotamia, Europe or to the east via land and sea (Ben-Yehoshua et al., 2012). In fact, the period spanning from the third century BCE to the second century CE represents the golden era for frankincense trade (Ben-Yehoshua et al., 2012; Groom, 1981).

Much archaeological evidence has been found to indicate trading and various uses of incense by ancient civilizations. The ancient Sumer used (*SIM*, *GIG*) as a code for 'incense', and a document dated 2300–2100 BCE states that 'amounts of incense resin in bags arrived by boat' (Steinkeller & Postgate, 1992). Various Assyrian and Babylonian documents included official trading lists, including codes translated as frankincense such as '*kanaktum*' at approximately 2300 BCE, '*I. GIV Ka.-Na-AK-tim*' at 1768 BCE and '*SIM.HI.A*' at approximately 1000 BCE (Campbell & Thompson, 1949; Snell, 1982). Many of these documents recorded frankincense/incense importation by Mesopotamia and eastern Arabia (Delmont). Later Assyrian documents from the fifth century BCE confirm attempts to control the origin of the frankincense trade in southern Arabia (Ben-Yehoshua et al., 2012). Hence, that region was also the most likely source of this valued commodity during earlier centuries. In addition, there is much archaeological evidence to confirm the import and use of myrrh and frankincense by ancient Egyptians. Several records document expeditions to import this valued commodity from the Land of Punt, which according to a recent investigation spread from northern Somalia to Eretria, eastern Ethiopia and part of Sudan. The earliest record goes back to 2800 BCE during the time of King Sahure, and later depictions took place during the reign of Queen Hatshepsut (1473–1458 BCE) (Ben-Yehoshua et al., 2012).

Religious and Cultural Uses

Ancient Egyptians, Mesopotamians, Greeks, Romans, Jews, Christians and Muslims used frankincense in their religious ceremonies and in various social and cultural rituals and events associated with childbirth, weddings and funerals from the third millennium BCE, and their modern counterparts continue to use it at present.

The ancient Egyptians were documented to have burned frankincense and myrrh as part of their religious and cultural ceremonies. Incense smoke was believed to carry prayers upwards to gods, and idols were regularly scented with incense (Ziegler, 1932). The large temples of that age contained places to store and handle this vital commodity (Brun, 2000). As part of their cultural traditions, the Egyptians also scented their homes with incense on a regular basis (Ziegler, 1932).

Babylon was recorded to have offered 1000 talents (45 tons) of frankincense during their Bel festival each year. The Greeks and Romans were noted for burning multiple tons of frankincense during some of their public ceremonies (Coder, 2011). The development of a wealthy middle class throughout the Roman Empire led to a high demand and use of exotic products that included frankincense not only for religious rituals but also as part of personal care, scenting, purification and the display of status and material success (Ben-Yehoshua et al., 2012).

Both the Old Testament and the New Testament contain several verses that extoll myrrh and frankincense for their sweet odours, which induce spiritual excitement and rich sensations of love and provide tranquillity. For that reason, frankincense was used to accompany prayers, and both Judaic and Christian faiths have developed detailed instructions for its use in various religious rituals (Ben-Yehoshua et al., 2012).

Although Islam does not require the use of incense in religious activities, Muslims have maintained the use of incense traditions in scenting their homes, mosques, public gathering halls, clothes and bodies over the centuries. Among the indirect benefits of incense is its ability to serve as an insect repellent against mosquitoes and flies.

In Oman, as well as other places in the Arabian Peninsula, frankincense is used to symbolize the welcoming of guests into a house or at the end of a reception and to accompany wedding ceremonies to spread pleasant, warm feelings. Even the body of a dead person is scented with frankincense during its final preparation prior to burial.

Cosmetic Use

Throughout the long history of human frankincense use, it is evident that people have explored its potential use beyond the sweet odour of its white smoke. Among its early cosmetic uses was the production of perfumed oils of frankincense and other resins, especially at Petra by the Nabatean as early as the second century BCE. This use allowed the Nabateans to change their role from simple middlemen in the incense trade to producers of new innovative products that utilized frankincense and other resins as primary ingredients. Hence, pure oils were further diluted or mixed with other oils that were extracted from locally available plants to maximize the added value upon trading and export to the major cities of that time. The ceramic containers used for these shipments were made especially for that purpose in Petra itself (Johnson, 1987). The popularity of public baths throughout the Roman Empire required the major use of perfumed oils. Perfumed oils extracted from frankincense continued to be used into modern time by perfumers as an absolute, especially in Oriental bases, since it blends well with various natural aromas such as musk, amber and floral derivatives.

Ancient Egyptians also produced *kohl*, a black eyeliner that was used by both women and men and produced from the soot of burned incense, which is a tradition that continues into the modern era, not only in Egypt but also in other parts of Arabia. This product and other beauty unguents involving fragrant resins such as frankincense were in universal use and were highly prized by the Greeks. They continue to be used in modern-day cosmetics such as in soaps, creams and powders involved in skin care and aromatherapy. In Dhofar, women continue to use frankincense in many beauty formulations, including hair-smoothing treatments. Even the bark of the frankincense tree is used in tanning and dye preparations, primarily to dye cotton-based clothes and leather.

Medicinal Use

Ancient Egyptians used frankincense for various medical treatments, according to the *Ebers Papyrus* dated 1500 BCE (Ben-Yehoshua et al., 2012). These included treatments for throat and larynx infections, vomiting control, phlegm reduction, reducing asthmatic attacks and stopping bleeding. As early as the sixth century CE, the ancient Chinese were documented to have used frankincense in various treatments, including treatments for enhanced blood circulation, the relief of indigestion and wound and injury treatments. The Ayurvedic medical system in India prescribed frankincense gum for arthritic and inflammatory cases, pulmonary diseases and gastric conditions (Miller & Morris, 1988). Ibn Sina (Avicenna) prescribed frankincense for dysentery, ulcers, fevers, vomiting and tumours (Jahier & Noureddine, 1956). This traditional medical use continues with frankincense used as a household natural or herbal treatment against coughs, throat swelling, dental problems and wounds.

Some commercially available frankincense-based ointments were intended to reduce arthritic conditions, while others are reported to be effective in wound treatments (Ben-Yehoshua et al., 2012). Numerous patents have been filled in the USA and Europe since 1997 based on frankincense or one or more compounds of its derivatives. The medical treatment potential of these products include anti-inflammatory properties, the treatment of acute respiratory distress syndrome, the treatment of neoplasia and the prevention or treatment of cerebral ischaemia, cerebral lesions, Alzheimer's disease, brain tumours and other cancers (Ben-Yehoshua et al., 2012).

A review of recent studies on frankincense extracts or derivatives of pure compounds confirmed their therapeutic potential in vitro, in vivo and through limited clinical trials in a variety of diseases and disorders (Mckenna & Hughes, 2014). These uses include its anti-inflammatory role in treating rheumatoid arthritis; the role of boswellic acids, especially acetyl-11-keto-β-boswellic acid (AKBA), as an anti-inflammatory against intestinal inflammation; its antimicrobial effects against the human pathogens *Candida albicans*, *Staphylococcus epidermidis* and *S. aureus*; the inhibitory effects of AKBA on tumour growth; and the direct effect of incensole acetate, as one of its biomarker derivatives, on the central nervous system because it induces a sensation of 'warmth' and works as an antidepression factor (Mckenna & Hughes, 2014).

References

Ben-Yehoshua, S., Borowitz, C., & Ondrej Hanuš, L. (2012). 1 frankincense, myrrh, and balm of Gilead: Ancient spices of Southern Arabia and Judea. *Horticultural Reviews, 39*(1), 3–66.

Bowen, R. (1958). Ancient trade routes in South Arabia. *Archaeological Discoveries in South Arabia*, 35–42.

Brun, J.-P. (2000). The production of perfumes in antiquity: The cases of Delos and Paestum. *American Journal of Archaeology, 104*, 277–308.

Campbell, J., & Thompson, R. (1949). *A dictionary of Assyrian Botany.* Paper presented at the British Academy, London.

Coder, K. D. (2011). Frankincense & myrrh: A gift of tree history. University of Georgia, Georgia, USA.

Coppen, J. J. (1995). Gums, resins and latexes of plant origin. Non-wood forest products 6. *Rome: Food and Agriculture Organization of the United Nations x, 142p. ISBN, 661102982.*

Coppi, A., Cecchi, L., Selvi, F., & Raffaelli, M. (2010). The frankincense tree (*Boswellia sacra*, Burseraceae) from Oman: ITS and ISSR analyses of genetic diversity and implications for conservation. *Genetic Resources and Crop Evolution, 57*(7), 1041–1052.

Ebeling, E. (1948). Mittelassyrische Rezepte zur Bereitung von wohlriechenden Salben. *Orientalia, 17*(2), 129–145.

Ebeling, E. (1949). Mittelassyrische Rezepte zur Bereitung von wohlriechenden Salben (Fortsetzung). *Orientalia, 18*(4), 404–418.

Giuliani, C., Bini, L. M., & Lippi, M. M. (2012). Two structures and functions in the nectary of frankincense tree (*Boswellia sacra* Flueck.). *Flora-Morphology, Distribution, Functional Ecology of Plants, 207*(1), 74–79.

Glueck, N. (1940). *The other side of the Jordan*: American schools of oriental research.

Grohmann, A. (1933). *Südarabien als Wirtschaftsgebiet* (Vol. 4). Forschungsinstitut für Osten und Orient.

Groom, N. (1981). Frankincense and myrrh. *A study of the Arabian incense trade. Longman: London & New York, 285*, 96–120.

Hepper, F. N. (1969). Arabian and African frankincense trees. *The Journal of Egyptian Archaeology, 55*(1), 66–72.

Jahier, H., & Noureddine, A. (1956). Les belles lettres. In *Avicenne, Poème de la médecine*. Paris: Société d'Édition.

Johnson, D. J. (1987). *Nabataean trade: intensification and culture change*: University of Utah. Ann Arbor, Univesity Microfilms, Michigan, USA.

Mckenna, D. J., & Hughes, K. (2014). *The incense bible: Plant scents that transcend world culture, medicine, and spirituality*: Routledge. New York, USA.

Miller, A. G., & Morris, M. (1988). Plants of Dhofar: the southern region of Oman, traditional, economic and medicinal uses. *Oman: Office of the Adviser for Conservation of the Environment, Diwan of Royal Court Sultanate of Oman xxvii, 361p.-col. illus.. ISBN, 715708082*.

Raffaelli, M., Mosti, S., & Tardelli, M. (2003). The frankincense tree (*Boswellia sacra* Flueck., Burseraceae) in Dhofar, southern Oman: Field-investigations on the natural populations. *Webbia, 58*(1), 133–149.

Snell, D. C. (1982). *Ledgers and prices: Early Mesopotamian merchant accounts*: Yale University Press, Connecticut, USA.

Steinkeller, P., & Postgate, J. N. (1992). *Third-millennium legal and administrative texts in the Iraq Museum, Baghdad* (Vol. 4). Winona Lake: Eisenbrauns.

Thulin, M., & Warfa, A. M. (1987). The frankincense trees (Boswellia spp., Burseraceae) of northern Somalia and southern Arabia. *Kew Bulletin, 42*, 487–500.

Tolera, M., Sass-Klaassen, U., Eshete, A., Bongers, F., & Sterck, F. J. (2013). Frankincense tree recruitment failed over the past half century. *Forest Ecology and Management, 304*, 65–72.

Van Beek, G. W. (1958). Frankincense and myrrh in ancient South Arabia. *Journal of the American Oriental Society, 78*(3), 141–152.

Van Beek, G. W. (1960). Frankincense and myrrh. *The Biblical Archaeologist, 23*(3), 70–95.

Ziegler, G. M. (1932). The diuturnal use of perfumes and cosmetics. *The Scientific Monthly, 34*(3), 222–237.

Chapter 2
Taxonomy, Distribution and Ecology of *Boswellia*

Burseraceae Family

The Burseraceae family consists of 19 genera and more than 700 species of plants (Doyle & Hotton, 1991; Ii, 2003). This family includes many shrubs and trees that are distributed throughout Africa, the Middle East, the Indian subcontinent and the Americas (Table 2.1). This family has three subtribes, Bursereae (Burserinae and Boswelliinae), Canarieae and Protieae (Group, 2009). This family has been considered as a sister group of the Anacardiaceae (Gadek et al., 1996; Weeks, Daly, & Simpson, 2005). In Protieae, *Protium* consists of approximately 147 species; *Commiphora* has approximately 200 species in the Bursereae tribe and *Canarium* comprises 75 species in the tribe Canarieae (Group, 2009).

All the taxa within the Canarieae plus the Boswelliinae clade lack pseudarils. Both *Boswellia* and *Triomma* have dry, schizocarpic fruits. *Garuga* and *Boswellia* have similarly shaped prolate spheroidal pollen, a finely perforate tectum and pronounced endexine thickenings in the pollen wall (Harley & Daly, 1995). The Canarieae plus Boswelliinae clade may have evolved from a Southeast Asian ancestor (Gadek et al., 1996). However, the basal species of this clade (*Boswellia*, *Garuga*, *Triomma*) are African, Indian or Southeast Asian in distribution (Gadek et al., 1996). The sole study on Sapindales molecular phylogeny showed the least support for basal relationships among the three major clades that make up the family, namely, Canarieae plus Boswelliinae, the Protieae and the Burserinae (Gadek et al., 1996; Weeks et al., 2005). Effort is needed to elucidate the Burseraceae with respect to genotypes and phenotypes (Clarkson, Chase, & Harley, 2002; Mahekar & Yadav, 2006). According to Weeks et al. (2005), the phylogenetic analysis showed that *Boswellia* forms a cladogram with *Garuga* through old-world distribution, whereas *Bursera* and *Commiphora* are forming closer clades. There have been various attempts to elucidate the Sapindales and Burseraceae, in which mostly *Commiphora*, *Bursera*, *Canarium* and *Protium* have been studied to some extent, but *Boswellia* has been ignored and is the least understood (Gadek et al., 1996; Weeks et al., 2005).

A. Al-Harrasi et al., *Biology of Genus Boswellia*,
https://doi.org/10.1007/978-3-030-16725-7_2

Table 2.1 List of genera in the Burseraceae, approximate number of species, tribal and subtribal placement and geographic range

Subtribes	Genus	Number of species	Location
Burserinae	*Bursera Jacq.* ca.	100 spp.	Caribbean, Mexico, Central and S. America
	Commiphora Jacq. ca.	190 spp.	Africa, India, S. America
Boswelliinae	*Aucoumea Pierre*	1 sp.	W. Africa
	Beiselia Forman	1 sp.	SW Mexico
	Boswellia Roxb. ca.	30 spp.	NE Africa, Arabia, India
	Triomma Hook. f.	1 sp.	W. Malesian region
	Garuga Roxb	4 spp.	India, SE Asia
Canarieae	*Canarium L.* ca.	105 spp.	SE Asia, Malaysia, Africa
	Dacryodes Vahl	66 spp.	Caribbean, Mexico, C. and S. America, SE Asia, Africa
	Haplolobus H.J. Lam	22 spp.	E. Malaysia
	Pseudodacryodes R. Pierlot	1 sp.	Central Africa
	Rosselia Forman	1 sp.	Rossel Island, New Guinea
	Santiria Blume	24 spp.	W. Malesian region, Philippines, Moluccas, New Guinea, Africa
	Scutinanthe Thwaites	2 spp.	Sri Lanka, S. Myanmar, Celebes,Sumatra, Malay Peninsula, Borneo
	Trattinnickia Willd	13 spp.	C. and S. America
	Crepidospermum Hook. f.	6 spp.	S. America
Protieae	*Protium Burm. f.*	150 spp.	Mexico, C. and S. America, Africa, SE Asia
	Tetragastris Gaertn	9 spp.	Central and S. America

Adopted from Weeks et al. (2005)
Garuga was placed informally within the Bursereae by Harley and Daly (1995)

Further studies are essential to understand not only the taxonomy of Burseraceae but also the *Boswellia* genus.

In other classifications, the Bursereae comprises *Bursera*, *Commiphora*, *Aucoumea*, *Beiselia*, *Boswellia*, *Triomma*, *Garuga*, *Ambilobea*, *Canarium*, *Dacryodes*, *Haplolobus*, *Pseudodacryodes*, *Rossellia*, *Santiria*, *Scutinanthe*, *Trattinnickia*, *Crepidospermum*, *Protium* and *Tetragastris* (De-Nova et al., 2012; Lawrence, 1951).

Table 2.2 *Boswellia* species and their distribution

1	*Boswellia sacra*	Oman, Yemen
2	*Boswellia serrata*/*Boswellia thurifera*/*Boswellia glabra*	India
3	*Boswellia ovalifoliolata*	India
4	*Boswellia pirottae*	Ethiopia
5	*Boswellia carterii*	Somalia
6	*Boswellia frereana*	Somalia
7	*Boswellia globose*	Somalia
8	*Boswellia ogadensis*	Ethiopia
9	*Boswellia neglecta*/*Boswellia elegans*/*Boswellia hildebrandtii*/*Boswellia microphylla*/*Boswellia campestris*/*Boswellia holstii*/*Boswellia multifoliolata*	Ethiopia, Somalia, Kenya
10	*Boswellia rivae*/*Boswellia boranensis*/*Boswellia ruspoliana*	Ethiopia, Somalia, Kenya
11	*Boswellia papyrifera*/*Boswellia chariensis*/*Boswellia occidentalis*	Ethiopia, Eritrea, Sudan
12	*Boswellia odorata*	Niger, Northern Nigeria and eastern Cameroon
13	*Boswellia dalzielii*	Northern Nigeria
14	*Boswellia madagascariensis*	Madagascar
15	*Boswellia popoviana*	Socotra
16	*Boswellia nana*	Socotra
17	*Boswellia bullata*	Socotra
18	*Boswellia dioscorides*	Socotra
19	*Boswellia ameero*	Socotra
20	*Boswellia elongata*	Socotra
21	*Boswellia socotrana*	Socotra

Boswellia madagascariensis has recently been given its own genus, *Ambilobea* (Thulin, 2004), and *Boswellia bricchettii* is synonymous with *Lannea obovate* and is thus not a *Boswellia* species

Boswellia Genus

The fragrant resin-producing tree is placed in the genus *Boswellia*, which was named after the botanist John Boswell in 1846. It is a member of the Burseraceae-flowering plant family, which includes at least 16 other genera. This family is known as the 'torch wood family' for the quality of light from the burned wood of its members. It is endemic to Africa, Asia and the Americas and characterized by latex, oils or resin formation, and its aromatic bark tends to peel off (Group, 2009; Judd, Campbell, Kellogg, & Stevens, 1999). There are 19 described species in the genus *Boswellia*, and there are several varieties, sub-species and forms within a few of the species. Many of the species have been described multiple times with different names, which led to a taxonomic controversy in the literature. All the true *Boswellia* species described thus far are as follows (Table 2.2):

Historical Perspectives on *Boswellia* Taxonomy

The historical aspects of frankincense and its trade routes were discussed in Chap. 1. This section summarizes the historical background of *Boswellia* taxonomy. The initial work of Howes (F. Howes, 1950; F. N. Howes 1929; Frank Norman Howes, 1949) highlighted the importance of the frankincense obtained from three species. Van Beek (Van Beek, 1960) has described the morphology of the tree and its ability to produce frankincense. Hepper (1969) showed that there are three species in Somalia, viz. *B. bhau-dajiana*, *B. carterii* and *B. frereana*. His report was the first to present a detailed nomenclature of the three species and their morphological differentiation. Groom (1981) also provided a close look at the frankincense-producing tree and distinguished it from the myrrh-producing tree. A major contribution was made by Tucker (1986) and Thulin and Warfa (1987) in understanding the taxonomic distribution of *Boswellia*. While Tucker (1986) agreed to the taxonomic differentiation between four species of *Boswellia* as proposed by Hepper, Thulin and Warfa (1987) disagreed with this taxonomy. Thulin and Warfa (1987) rejected the whole literature on frankincense-producing trees and suggested that only *B. sacra* and *B. frereana* are essential for producing resins of commercial value. In addition, the authors also indicated that *B. papyrifera* is closer to *B. sacra*. This classification was later separated into separate species by Rees (1995). *B. serrata* was further described by Rees (1995) in India, and *B. ameero* and *B. socotrana* were from the Socotra Island in Yemen. Doyle and Hotton (1991) explained the characteristics of pollens from *Boswellia* during their analysis of angiosperms. Gadek et al. (1996) have used the genetic information from three species of *Boswellia* to understand the molecular delimitations and groupings of Spindales. Similarly, APG-II (2003) explained the family of *Boswellia* through a phylogenetic analysis of angiosperms. Mertens, Buettner and Kirchhoff (2009) assessed the volatile components of *B. sacra*, *B. frereana*, *B. serrata*, *B. papyrifera*, *B. rivae* and *B. neglecta* and found more than 300 volatiles in frankincense. The authors showed variable concentrations of major constituents in the frankincense. *B. sacra* and *B. carterii* have been suggested to be two different species on the basis of differences in the chemical composition of their chemotaxonomic markers, but Mertens et al. (2009) accounted for this variation as being due to geography but not because of its botanical classification. In this book, we adopt this idea that *B. sacra* and *B. carterii* will be considered as one but differ in their geographical locations (Oman for *B. sacra* and Somalia for *B. carterii*). Van Vuuren, Kamatou and Viljoen (2010) showed the taxonomic differentiation of five species of *Boswellia* in terms of their essential oil composition and variation across different samples produced from each species. This classification was correlated with the diversification in the contents of various monoterpenes (α-pinene, β-pinene, α-thujene, sabinene, myrcene, limonene and p-cymene). Recently, J Eslamieh (2011) described all the species of *Boswellia*. Niebler and Buettner (2016) reported 216 chemical constituents from the resins of 4 *Boswellia* species (*B. sacra*, *B. serrata*, *B. papyrifera* and *B. frereana*) and suggested a clear demarcation and discrimination among species based on the variations in chemodiversity. In another study, Niebler, Eslamieh and Buettner (2016) reported the volatile constituents of oils from the resins of *B. ameero*, *B. elongata*, *B. neglecta*, *B. popoviana* and *B. rivae*. The authors concluded that a chemical analysis of these species was not sufficient to allow for clear discrimination. The authors suggest that further studies with higher number of samples may support this differentiation. Recently, Al-Harrasi et al. (2018) published a book on the chemistry and biological activity of boswellic acids, in which a detailed description of the differentiation using the chemodiversity was presented.

Boswellia Species Description

The taxonomic work on *Boswellia* by Thulin and Warfa (1987) and later by Eslamieh (J Eslamieh, 2011) in particular includes detailed cultivation methods. The works of Jason Eslamieh (2010) are considered major contributions in this field. Thulin and Warfa (1987) presented a full botanical description of two new species from Socotra; however, the literature in this field is scarce. The leaf parts of most of the species are imparipinnate, but both subsessile and subalternate leaflets have been observed. The leaves are pubescent, but some of the smaller leaves are glabrous in nature. All species are without thorns; they are shrubby, small-to-medium trees, and the external bark is exfoliated. According to Jason Eslamieh (2010) and Thulin and Warfa (1987), the flowers are five-lobed and have five petals, with internal stamens that are disk-shaped in the centre. The fruits and seeds are similar to capsules, and the outer layer of three-to-five-winged pyrenes falls at maturity. These plant species can self-pollinate and are bisexual.

The frankincense tree is a deciduous plant known to grow on rocky terrains, solid rocks and gravel soils. In Dhofar and Somalia, the tree habitats are heavy and calcareous with rich limestone. The tree reaches 2–8 m in height. Normally, it branches close to the ground, giving the appearance of multi trunks, each up to 25 cm in diameter. Its bark is easily peeled off and can be stripped in long sheets that were once used for writing. The outer layer of the bark (periderm) makes coarse and loose papery-like flakes that peel off. In addition, the inner layer (cortex) is thick, rust or amber-coloured and measures approximately 2.5 cm thick. It has long, compound leaves with 7–9 small toothy leaflets that appear dark glossy green on the upper surface and light-hued on the underside. This species has star-shaped, very small white-yellowish flowers with rosy pink centres that are produced in axillary clusters. Each flower is composed of five petals and ten stamens on a single cup (Coder, 2011; Hepper, 1969; Vaishnav & Janghel 2019). The showy ring within the flower, which changes colour from yellow to rosy pink or brilliant red, is assumed to be a nectary ring and contains nectarostomata and glandular trichomes that attract pollinators, such as the honeybee (Hepper, 1969). Perhaps some of the frankincense-flavoured honey produced in Dhofar is an evidence of these interactions.

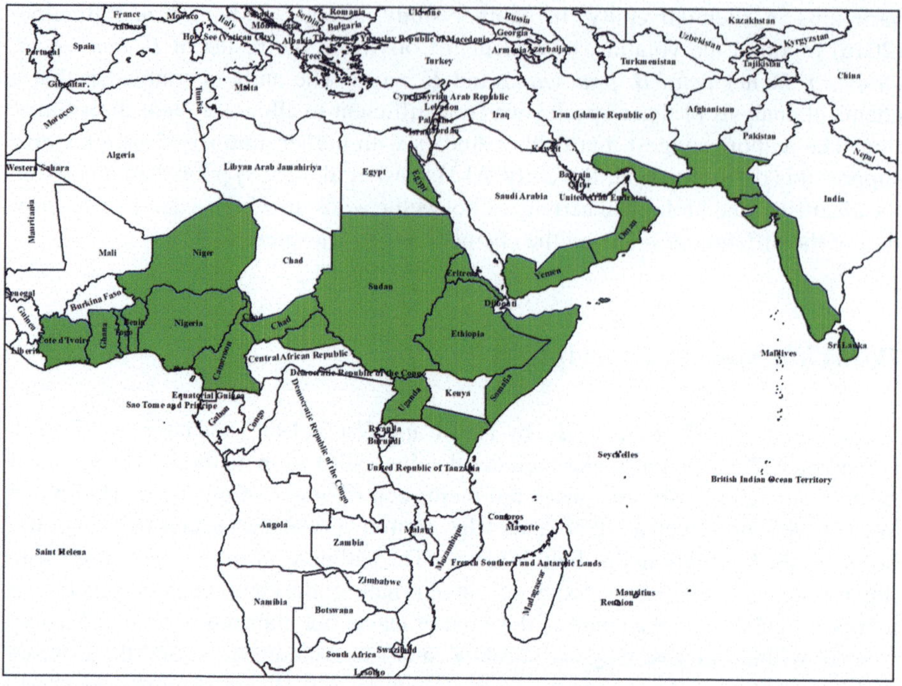

Fig. 2.1 Distribution of *Boswellia* species across different countries

Distribution of *Boswellia* Throughout the World

Almost all the known *Boswellia* species are found in the tropics of Africa, Yemen (including the Socotra Island), Oman, Iran, Pakistan and India (Fig. 2.1; Table 2.1).

Boswellia sacra Flueck

The *B. sacra* tree is a late successional species that normally pollinates under extreme environmental conditions, and it has been confronted with population decline (Lippi, Giuliani, Gonnelli, & Bini, 2011; Sunnichan, Mohan Ram, & Shivanna, 2005). The somatic chromosome number for *B. sacra* is $2n = 22$ (Thulin & Warfa, 1987). The pollen size is large, and the shape is sub-prolate to prolate (Daly, Harley, Martínez-Habibe, & Weeks, 2010). The tree is 1–2 m tall, and the branches often start from just above the aboveground portion; the bark is pale brown and peels off the tree similar to paper (Fig. 2.2, 2.3, 2.4, 2.5, 2.6 and 2.7). However, in some locations, this plant is more often found on cliffs and is sedentary on the ground. Its leaflets are lanceolate and foliolate (Fig. 2.2).

Fig. 2.2 *Boswellia sacra* leaf morphology. The copyrights for all the pictures in this chapter belong to the authors, unless otherwise indicated

Fig. 2.3 *Boswellia sacra* seed and flower morphology

Boswellia ameero Balf. f.

B. ameero is endemic to Socotra Island in Yemen, and it grows in dry and semi-deciduous woodlands at altitudes from 120 to 1050 m above sea level, mostly on rocky cliffs that are easily recognizable by their not fully bright pink flowers. The tree height ranges from 5 to 8 m, the trunk is grey, and the outer bark is yellowish (Fig. 2.8). The leaves appear dark green from above and below, and the size ranges from 14 to 25 cm. The flowers are red to pink with seven petals and are glabrous in nature.

Fig. 2.4 *Boswellia sacra* tree

Fig. 2.5 A *Boswellia sacra* farm (Wadi Dhoka National Park)

Fig. 2.6 A wild *Boswellia sacra* tree in Sidab (Dhofar, Oman)

Fig. 2.7 A wild *Boswellia sacra* tree in the Hasik area (Dhofar, Oman)

Fig. 2.8 *Boswellia ameero* (© Joël Lodé). The copyright permissions to publish these pictures have been kindly granted by Joel Lode

Boswellia dioscoridis Thulin

B. dioscoridis is endemic to Socotra Island in Yemen and is 3.5–6 m tall (Fig. 2.9). It grows on limestone cliffs in dry, semi-deciduous woodland in southern and eastern Socotra Island. Its leaflets are densely populated by shoot apices and are subsessile (Fig. 2.10). There are many flowers with leaves that are clustered, with panicles or racemes; the petals are white and pinkish, whereas the fruits are 3–5 locular and narrow at both ends (Fig. 2.9).

Fig. 2.9 *Boswellia dioscoridis* (© Joël Lodé). The copyright permissions to publish these pictures have been kindly granted by Joel Lode

Fig. 2.10 *Boswellia dioscoridis* (© Joël Lodé). The copyright permissions to publish these pictures have been kindly granted by Joel Lode

Boswellia popoviana Hepper

B. popoviana is endemic to Socotra Island in Yemen and has a scattered tree density on limestone-rich cliffs in dry, semi-deciduous woodlands, and it grows at altitudes ranging from 20 to 1050 m (Fig. 2.11). Thulin and Al-Gifri (1998) described the distribution of *B. popoviana*. The leaflets are either regularly crenate or deeply lobbed and irregularly crenate sinuate (Fig. 2.12).

Boswellia nana Hepper

B. nana is endemic to Socotra Island in Yemen and is recognized as a complex of low, sometimes totally prostrate shrubs and small trees found on limestone cliffs and areas of flat limestone pavement in a relatively restricted area of northeast

Fig. 2.11 *Boswellia popoviana* (© Joël Lodé). The copyright permissions to publish these pictures have been kindly granted by Joel Lode

Fig. 2.12 *Boswellia popoviana.* The copyright permissions to publish these pictures belong to the authors of this work

Socotra. Thulin and Al-Gifri (1998) regarded *B. nana* as a reduced form of *B. popoviana*. However, it differs from that species in that it has smaller, more reduced panicles of red, not yellowish, flowers, and the undersurface of its leaves are glabrous and not white-felted. Certain plants from inland cliffs in northeast Socotra, including *B. nana*, show significant variation in the degree of division in their leaves and in their habitat. Researchers strongly suspect that these plants are a result of a hybridization with *B. socotrana*. The clearest evidence for this hypothesis is a tree found at the base of limestone cliffs at Hamadero that is somewhat intermediate in appearance between *B. socotrana* and 'typical' *B. nana*. It bears both simple and pinnate leaves with the pinnate leaves having the winged rachises that are otherwise diagnostic of *B. socotrana*. This example also seems to be intermediate in its ecology; typical *B. nana* is restricted to cliffs, and *B. socotrana* is ground-rooted, whereas the intermediate plant was found growing among rocks at the base of cliffs

Fig. 2.13 *Boswellia nana* (© Joël Lodé). The copyright permissions to publish these pictures have been kindly granted by Joel Lode

(Fig. 2.13). Populations of *B. nana* were found on the seaward-facing escarpments of the north-eastern Socotra, and they displayed only simple leaves. These trees are growing on limestone cliffs and pavements, in the dry, semi-deciduous woodlands on the north-eastern part of the island. *Boswellia nana* grows at an altitude of 300–550 m as described and published in *Hooker's Icones Plantarum*.

Boswellia elongata Balf. f

B. elongata is endemic to Socotra Island in Yemen. Its natural habitats are subtropical or tropical dry forests and subtropical or tropical dry shrubland. Young leaves turn pinnate with age and are 15–30 cm long and grey green, dark green to purple grey with pale undersides, strongly incised and elongated (Fig. 2.14). Adult plants with simple leaves are possibly hybrids with *B. popoviana*. The tree stems reach 6–8 m tall, trunk 0.8–1 m in diameter with a strongly flaking yellowish, pale brown or bronze, peeling bark. The branches are thick and swollen, spreading more or less in a horizontal and wavy fashion. During cultivation under shelter, the trees usually do not exceed 2.20 m. As one of seven endemic species found on Socotra, it is a

Fig. 2.14 *Boswellia elongata* (© Joël Lodé). The copyright permissions to publish these pictures have been kindly granted by Joel Lode

medium to large deciduous pachycaul tree reaching 6–8 m tall with diverse forms and intricate webs of branching, appearing as one of the very distinctive plants in Socotra. The long, slender leaves are simple when young and turn pinnate and feathery with age. New growth has an intricate lacy texture similar to a spider web. The colour of the leaves varies from grey, to purple, to light green with pale undersides, depending on the environment.

Boswellia socotrana Balf. f.

B. socotrana is endemic to Socotra, Yemen. Its natural habitats are subtropical or tropical dry forests and subtropical or tropical dry shrubland. *B. socotrana* is a rare species, with a very limited area of occupancy; several populations show no sign of regeneration, suggesting a decline in the quality of their habitat. The tree is ground-dwelling, and it is readily distinguished from all the other *Boswellia* on the island by the leaves, which have winged stalks and numerous small leaflets as well as inconspicuous pale-yellow flowers that are borne on short, sparse panicles (Fig. 2.15).

Boswellia serrata Roxb. ex Colebr

B. serrata (Salai Guggul or Indian frankincense) grows in the mountainous regions of India and Pakistan. It is synonymous with *B. glabra* Roxb (https://plants.usda.gov/core/profile?symbol=BOSE5). This is a deciduous tree ranging from 5 to 20 m in height. The bark is yellowish-white with dark blotches. The bark, similar to that of other *Boswellia* species, exfoliates in thin, papery, smooth flakes. The leaves are also imparipinnate, arranged in an alternating manner and apically clustered, whereas the leaflets can range from 15 to 31 cm (Fig. 2.16). The flowers are bisexual, small and white. The flowers are arranged in axillary or subterminal fascicled racemes with five to seven lobes and petals, and they are white in colour (Sasidharan, Kerala Forest Research Institute, https://indiabiodiversity.org/species/show/31115). The general habitat is occasionally in the hills on the lower dry deciduous slopes to 900 m, in gravelly, poor soils. These trees are mostly distributed throughout central and peninsular India, whereas in Pakistan, some of the population has been reported in Baluchistan.

Fig. 2.15 *Boswellia socotrana*. The copyright permissions to publish these pictures belong to the authors of this work

Fig. 2.16 *Boswellia serrata*

Boswellia ovalifoliolata Balakr & A.N. Henry

Boswellia ovalifoliolata is found in India, in Andhra Pradesh state, in the tropical to subtropical dry deciduous forests of the peninsular part of India. This species has a generation length of 25 years. Saha, Thavasi and Jayalakshmi (2008) and Raju, Lakshmi, Ramana and Chandra (2012) reported that *B. ovalifoliolata* occurs in the foothills of the Seshachalam Hill ranges of the Eastern Ghats in the Chittoor, Cuddapah and Kurnool districts of Andhra Pradesh up to an altitude of approximately 600–900 m. According to the detailed taxonomy of *B. ovalifoliolata* by Raju et al. (2012), the leafy part is shaded from December to February, whereas flowering occurs from March to April (Fig. 2.17). The flowers offer both pollen and nectar. The seeds are winged, papery, compressed, 7 mm long, 4 mm wide and 19.9 ± 3.1 mg in weight. The mature fruits dehisce and disseminate their lightweight, papery winged seeds with the aid of the wind. The windiness of the study site being provides the necessary driving force for the effective dispersal of the seeds away from the parent trees. Seed germination occurs after rainfall, but further growth depends on the soil water and the plant's nutritional status (Raju et al., 2012).

Boswellia rivae

This species is also known as qura or black incense. It is a spreading deciduous shrub or 3–6 m. The leaves are compound and measure approximately 7 cm with 7–10 pairs of hairy leaflets, each with small rounded teeth (Fig. 2.18). Flowers with five pink petals appear with the new leaves on several stalks up to 4 cm long, at the

Fig. 2.17 *Boswellia ovalifoliolata*

Fig. 2.18 *Boswellia rivae*

tips of thick branchlets. In Africa, it is found in *Acacia* and *Commiphora* wooded grasslands in red sandy to stony soils in the Dry Kolla agroclimatic zone of Sidamo, Bale and Harerge with altitudinal height ranging from 250 to 800 m. This species is also found in the Gamo Gofa region, although it is not as common there (http://www.worldagroforestry.org/usefultrees/pdflib/Boswellia_rivae_ETH.pdf).

Boswellia frereana Bird

B. frereana is native to northern Somalia. It is reportedly cultivated in Yemen. Locally, it is known as Dhidin or Maydi and Yigaar tree. According to a book by J Eslamieh (2011), the tree height reaches up to 8 m, with a swollen base at the tree trunk. The leaves are oblanceolate and subglabrous, with glandular hairs, and they are 10–30 cm long (Fig. 2.19). The flowers are generally glabrous to pubescent raceme-like thyrses. The resin is copious, milky and yellowish.

Fig. 2.19 *Boswellia frereana*

Fig. 2.20 *Boswellia neglecta*

Boswellia neglecta S. Moore

B. neglecta grows in Ethiopia, Kenya, Somalia, Tanzania and Uganda (Mokria et al., 2017). It has a strong, zesty, tea tree-like aroma, and it is effective at healing wounds and scars very quickly. In some areas, it is possible to distinguish between forms with many leaflets closely corresponding to *B. hildebrandtii* and *B. multifoliolata* and forms with few leaflets corresponding to *B. microphylla*, but the variation in large parts of Somalia is continuous. The type of *B. neglecta*, which has approximately 17–21 leaflets, is intermediate (Fig. 2.20). In Ethiopia, *B. neglecta* is predominantly found in the dry *Acacia-Commiphora* woodlands of the southern and south-eastern parts of the country. Frankincense produced from *B. neglecta* is known as 'Borena-type' (Mokria et al., 2017). The tree is a shrub that can reach up to 8 m tall and has dark grey bark, less peeling and straight young branches (J Eslamieh, 2011). According to J Eslamieh (2011), the leaves are sparsely tomentose, ranging from 2 to 11 cm long. The flowers are white or greenish to white in nature and contain 8 mm pedicels and ovate petals.

Boswellia dalzielii Hutch

This tree reaches 13 m high, and it is found in the wooded savannah. It has characteristically pale papery bark that is peeling and ragged, and it is sometimes locally abundant near pure stands. This species ranges from the Northern Ivory Coast to Nigeria and into Cameroun and Ubangi-Shari. According to Eslamieh (2017), the

Fig. 2.21 *Boswellia dalzielii*

Fig. 2.22 *Boswellia carterii*

trunk is yellow or grey with exfoliation. The leaves are arranged in alternate and imparipinnate ways. The leaves are glabrous green and approximately 20–45 cm long (Fig. 2.21). Small white flowers may appear, while the tree is leafless, and they are fragrant (Burkill, 1985).

Boswellia carterii Flueck

This distinctive cliff-dwelling tree is characterized by its pinnate leaves and crinkly, deeply irregular crenate-serrate leaflets. There is some variation within the species; plants from the single population on the eastern part of the island differ from trees from the west in having red petals and leaves with fewer from 9 to 13 cm, more regularly and less deeply crenate-serrated leaflets (Fig. 2.22).

Boswellia bullata Thul. & Gifri

This plant is endemic to Socotra, Yemen; it is a distinctive cliff-dwelling tree characterized by its pinnate leaves and crinkly, deeply irregular crenate-serrate leaflets. There is some variation; plants from a single population in the eastern part of the island differ from trees from the west in having red (not greenish-yellow) petals and leaves and more regularly and less deeply crenate-serrate leaflets (Miller, 2004). *Boswellia bullata* (Fig. 2.23) is on the IUCN Red List of Threatened Species (2004). The leaves are densely crowded at the shoot apex and are alternately arranged with lengths ranging from 7 to 18 cm. Flowers are produced with the leaves and are narrow, tomentose and clustered at the end. According to J Eslamieh (2011), the fruits are 4–5 locular, broadly pyriform and glabrous in nature.

Fig. 2.23 *Boswellia bullata*

Boswellia globosa Thul.

B. globosa grows in Somalia. It is a small tree, c. 2.5–4 m tall; the bark is purplish black and smooth; and the young shoots are greyish to purplish brown, longitudinally ridged and pubescent with spreading straight to crisped short hairs (Thulin, 2004). The leaves are (5–)11–21-foliolate, 2.5–6 cm long; petiole 2–12 mm long; rachis densely pubescent with spreading glandular hairs, not winged; and leaflets subsessile, opposite or sub-opposite, 3–6.5 × 1.5–4 mm, elliptical to obovate, and the inner ones are often suborbicular, obtuse at the apex, cuneate to rounded and ± asymmetric at the base, entire, somewhat paler beneath and densely pubescent with spreading hairs on both surfaces (Thulin, 2004). The flowers in the few-flowered pubescent racemes are up to c. 4 cm long; pedicels 2.5–7 mm long in fruit; and bracts triangular, 1–2 mm long, pubescent. Calyx pubescent, c. 1.2 mm long; petals c. 2 mm long, medially pubescent outside; filaments c. 0.8 mm long, ovate with filiform tip, papillose; anthers c. 0.6 mm long, papillose; style c. 0.8 mm long; and stigma truncate. Fruits 3–4-celled, subglobose, 8–12 × 7–11 mm, with wings up to 5 mm wide, glabrous; c. 6.5–7 × 5.5–6 mm, rugose, and surrounded by a persistent wing and forming a broadly ovate to sub-circular c. 8–12 mm long dispersal unit. The bark is used for tanning (Thulin 2004).

Boswellia pirottae Chiov

The tree height ranges up to 10 m with bark that is grey to black in colour and less peeling than in other species. The leaves are 15–35 cm long and glabrous and have pubescent petioles (1.5–4 cm; J Eslamieh (2011). The flowers are white, pubescent, narrow-contracted, and 2.5–10 cm long, with pubescent panicles, and clustered, and the petals are 3 mm long. This species is distributed from 1200 to 1800 m over the rocky slopes of Ethiopia (Reddy & Raju, 1999). Its wood is used for making paper and packing cases, toys and knife sheaths. It is endemic to Ethiopia.

B. papyrifera

B. papyrifera is distributed across the dry parts of Africa from Nigeria to Eritrea and Ethiopia (Ogbazghi) in highland tropical dry forests with shallow soils. The tree height reaches up to 12 m, whereas the leaves are pinnately compound and consist of 6–8 pairs of leaflets and one at the top. Flowering usually occurs during the dry season, when the leaves have fallen off. The flowers are red, and the flower stalk reaches up to 35 cm long, bearing white-pink flowers with five petals and ten yellow stamens. The bark is pale yellow brown, with papery peeling in wide strips. The slashed wood is reddish and exudes a fragrant resin (J Eslamieh, 2011). The seed is a red capsule that is three-sided, with three hard seeds, and it usually fruits during the wet season.

Boswellia microphylla Chiov

This species is a shrubby to 4-m-high tree, with dark grey to bluish grey bark. The leaves are 0.5–4.5 cm long and glabrous to sparsely pubescent. The flowers are white or creamy in colour with glabrous to sparely pubescent racemes. The petals are 2–3 mm long. The fruit is rounded to truncate at its apex (J Eslamieh, 2011). This species is distributed over Ethiopia, Somalia and Kenya, normally on gravelly soil with limestone and at a height of 400–1300 m (J Eslamieh, 2011).

References

Al-Harrasi, A., Rehman, N. U., Khan, A. L., Al-Broumi, M., Al-Amri, I., Hussain, J., … Csuk, R. (2018). Chemical, molecular and structural studies of Boswellia species: β-Boswellic Aldehyde and 3-epi-11β-Dihydroxy BA as precursors in biosynthesis of boswellic acids. *PLoS One, 13*(6), e0198666.

APG-II. (2003). *An update of the Angiosperm Phylogeny Group classification for the orders and families of Flowering plants:*. Paper presented at the APG II.

Burkill, H. (1985). The useful plants of West Africa Vol. 1. *Royal botanical gardens*, 386–387.
Clarkson, J. J., Chase, M. W., & Harley, M. M. (2002). Phylogenetic relationships in Burseraceae based on plastid rps16 intron sequences. *Kew Bulletin, 57*, 183–193.
Coder, K. D. (2011). *Frankincense & myrrh: A gift of tree history*. University of Georgia, Georgia, USA.
Daly, D., Harley, M., Martínez-Habibe, M., & Weeks, A. (2010). Burseraceae *Flowering Plants. Eudicots* (pp. 76–104). Springer.
De-Nova, J. A., Medina, R., Montero, J. C., Weeks, A., Rosell, J. A., Olson, M. E., … Magallón, S. (2012). Insights into the historical construction of species-rich Mesoamerican seasonally dry tropical forests: the diversification of Bursera (Burseraceae, Sapindales). *New Phytologist, 193*(1), 276–287.
Doyle, J. A., & Hotton, C. L. (1991). Diversification of early angiosperm pollen in a cladistic context. *Pollen and spores: patterns of diversification, 169*, 195.
Eslamieh, J. (2010). Creating "Perfect" Boswellia. *Cactus and Succulent Journal, 82*(3), 126–131.
Eslamieh, J. (2011). Cultivation of Boswellia Sacred Trees of Frankincense. *Phoenix (US): A Book's Mind. viii.*
Gadek, P. A., Fernando, E. S., Quinn, C. J., Hoot, S. B., Terrazas, T., Sheahan, M. C., & Chase, M. W. (1996). Sapindales: molecular delimitation and infraordinal groups. *American Journal of Botany, 83*(6), 802–811.
Groom, N. (1981). Frankincense and myrrh. A study of the Arabian incense trade. *Longman: London & New York, 285*, 96–120.
Group, A. P. (2009). An update of the Angiosperm Phylogeny Group classification for the orders and families of flowering plants: APG III. *Botanical Journal of the Linnean Society, 161*(2), 105–121.
Harley, M. M., & Daly, D. C. (1995). Burseraceae Kunth.: Protieae March. em. Engl. *World pollen and spore flora*, 20.
Hepper, F. N. (1969). Arabian and African frankincense trees. *The Journal of Egyptian Archaeology, 55*(1), 66–72.
Howes, F. (1950). Age-old resins of the Mediterranean region and their uses. *Economic Botany, 4*(4), 307–316.
Howes, F. N. (1929). Tapping Peru-balsam. *Kew bulletin*, 327–329.
Howes, F. N. (1949). *Vegetable gums and resins*. Waltham, MA: Chronica Botanica Company.
Ii, A. (2003). An update of the Angiosperm Phylogeny Group classification for the orders and families of flowering plants: APG II. *Botanical Journal of the Linnean Society, 141*(4), 399–436.
Judd, W., Campbell, C. S., Kellogg, E., & Stevens, P. (1999). *Plant systematics. A phylogenetic approach* (Vol. 464, pp. 3–4). Sunderland, MA: Sinauer Associates.
Lawrence, G. H. M. (1951). *Taxonomy of Vascular Plants*. New York, NY: The Macmillan Company.
Lippi, M. M., Giuliani, C., Gonnelli, T., & Bini, L. M. (2011). Floral color changes in Boswellia sacra Flueck.(Burseraceae): A dialogue between plant and pollinator. *Flora-Morphology, Distribution, Functional Ecology of Plants, 206*(9), 821–826.
Mahekar, P. D., & Yadav, S. (2006). *Medicinal Plant of South Western Maharashtra* (pp. 75–99). Biodiversity of India.
Mertens, M., Buettner, A., & Kirchhoff, E. (2009). The volatile constituents of frankincense–a review. *Flavour and Fragrance Journal, 24*(6), 279–300.
Miller, A. (2004). Boswellia elongata. [Press release]
Mokria, M., Tolera, M., Sterck, F. J., Gebrekirstos, A., Bongers, F., Decuyper, M., & Sass-Klaassen, U. (2017). The frankincense tree Boswellia neglecta reveals high potential for restoration of woodlands in the Horn of Africa. *Forest Ecology and Management, 385*, 16–24.
Niebler, J., & Buettner, A. (2016). Frankincense revisited, part I: comparative analysis of volatiles in commercially relevant Boswellia species. *Chemistry & Biodiversity, 13*(5), 613–629.
Niebler, J., Eslamieh, J., & Buettner, A. (2016). Frankincense Revisited, Part II: Volatiles in Rare Boswellia Species and Hybrids. *Chemistry & Biodiversity, 13*(5), 630–643.

Raju, A., Lakshmi, P. V., Ramana, K. V., & Chandra, P. H. (2012). Entomophily, ornithophily and anemochory in the self-incompatible Boswellia ovalifoliolata Bal. & Henry (Burseraceae), an endemic and endangered medicinally important tree species. *Journal of Threatened Taxa, 4*(7), 2673–2684.

Reddy, K., & Raju, R. V. (1999). Plants in ethnoveterinary practices in Anantapur district, Andhra Pradesh. *Journal of Economic and Taxonomic Botany, 23*(2), 347–357.

Rees, A. (1995). *Frankincense and myrrh. New Plantsman, 2* (pp. 55–59).

Saha, S., Thavasi, R., & Jayalakshmi, S. (2008). Phenazine pigments from Pseudomonas aeruginosa and their application as antibacterial agent and food colourants. *Research Journal of Microbiology, 3*(3), 122–128.

Sunnichan, V., Mohan Ram, H., & Shivanna, K. (2005). Reproductive biology of Boswellia serrata, the source of salai guggul, an important gum-resin. *Botanical Journal of the Linnean Society, 147*(1), 73–82.

Thulin, M. (2004). New species of Boswellia and Commiphora (Burseraceae) from Somalia. *Nordic Journal of Botany, 24*(4), 373–376.

Thulin, M., & Al-Gifri, A. (1998). The frankincense tree (Boswellia spp.) of Socotra. Paper presented at the Socotra—Proceedings of the first international symposium on Socotra Island: present and future. UNDP, New York.

Thulin, M., & Warfa, A. M. (1987). The frankincense trees (Boswellia spp., Burseraceae) of northern Somalia and southern Arabia. *Kew Bulletin, 42*, 487–500.

Tucker, A. O. (1986). Frankincense and myrrh. *Economic Botany, 40*(4), 425–433.

Vaishnav, V., & Janghel, U. (2019). A note on the clonal propagation of depleted threatened species Boswellia serrata Roxb. through branch cuttings.

Van Beek, G. W. (1960). Frankincense and myrrh. *The Biblical Archaeologist, 23*(3), 70–95.

Van Vuuren, S., Kamatou, G., & Viljoen, A. (2010). Volatile composition and antimicrobial activity of twenty commercial frankincense essential oil samples. *South African Journal of Botany, 76*(4), 686–691.

Weeks, A., Daly, D. C., & Simpson, B. B. (2005). The phylogenetic history and biogeography of the frankincense and myrrh family (Burseraceae) based on nuclear and chloroplast sequence data. *Molecular Phylogenetics and Evolution, 35*(1), 85–101.

Chapter 3
Frankincense: Tapping, Harvesting and Production

What Is Resin?

Plant resin, which is sometimes known as sap or exudate, has been defined as a mixture of volatile and non-volatile terpenoids and/or phenolic secondary metabolites (Langenheim, 2003). These compounds are insoluble in water and are secreted in specialized internal structures or near the surface of the plant. Most of these resin secretions make up part of the plant defence mechanisms or act in response to ecological interactions. Most of the resin products are subject to air-drying upon secretion, and they harden with the passage of time. These solid masses are either crystalline, turbid, of a single or mixture of colours and of variable sizes depending on the species and their environmental conditions (Fig. 3.1). However, resins are different from gums, mucilages, oils, waxes and latexes (Langenheim, 2003; Table 3.1). In addition, resins are also divided further into oleoresins, balsams and copals. Notably, resins can be hard (Taniguchi et al., 2014) or soft (Martín-Ramos, Fernández-Coppel, Ruíz-Potosme, & Martín-Gil, 2018). Similarly, the natural resin produced by *Bursera* is known as copal, the resin produced by *Boswellia* is frankincense, and that produced by *Commiphora* is myrrh (Eslamieh, 2010).

How Do the Trees Produce Resin?

The constituents of the resin are produced as a result of biosynthesis either through the shikimic acid or mevalonic acid pathway (Keeling & Bohlmann, 2006; Lange & Ghassemian, 2003) as phenolics or terpenes, respectively. The smooth endoplasmic reticulum (ER) and leucoplast are the known synthesis sites of the essential terpenes in the resin. Once synthesized in the epithelial cells, the resin is exported from vesicles, which are enveloped in a plasma membrane. Afterwards, the resin is discharged via turgor pressure into intercellular storage spaces (Boschiero &

A. Al-Harrasi et al., *Biology of Genus Boswellia*,
https://doi.org/10.1007/978-3-030-16725-7_3

Fig. 3.1 Frankincense from some *Boswellia* trees of different species

Tomazzello-Filho, 2012; Krokene & Nagy, 2012; Phillips & Croteau, 1999). These spaces can be resin sacs or secretory canals in the inner bark of *Boswellia* or other resin-producing trees such as *Commiphora*. The resin is synthesized through two strategies, namely, (i) constitutive, in which the resin is stored in canals, pockets, or sacs in the parenchyma regions or near the surface of the outer bark, and (ii) induced, in which it is formed by a reaction or in response to physical injury or tapping (Fig. 3.2). Since the parenchyma develops through the meristematic region (vascular cambium), the resin structures are either produced axially (vertically) or radially (horizontally). These canals are often interconnected, and the resin flow is dependent on the intensity of the wounds to the trees (Fig. 3.3). However, if needed, the plant signals related genetic transcripts to de novo synthesize more resin (Langenheim, 2003; Ogbazghi, Rijkers, Wessel, & Bongers, 2006; Ruel, Ayres, & Lorio, 1998; Wilkens, Ayres, Lorio, & Hodges, 1998). As part of the defence mechanism of the tree, resin production requires energy that is obtained through

Table 3.1 Properties and compositions of various secretory materials produced from plants

Type	Primary components	Solubility	Secretory tissue
Resins	Terpenoids; phenolic	Lipid-soluble	Canals, pockets, compound cavities, trichomes, epidermal cells
Gums	Polysaccharides	Water-soluble	Cavities
Mucilages	Polysaccharides	Water-soluble	Idioblasts, epidermal cells, trichomes, ducts, cavities
Oils (fats)	Fatty acids and glycerol	Lipid-soluble	None
Waxes	Fatty acids esterified with long-chain alcohols	Lipid-soluble	Unspecialized Epidermal cells
Waxes	Complex mixture, may include terpenoids, henolic compounds, roteins, carbohydrates, etc.	Lipid-soluble	Laticifers

Fig. 3.2 Tree tapping and the oozing of resin from a *B. sacra* tree

photosynthesis. Since this process is dependent on photosynthesis, it varies greatly from tree to tree depending on the available resources (water, light and nutrients) and the climatic conditions (drought, heat, etc.). If limited, all these growth conditions can limit the tree's ability to defend itself.

How Is the Resin Synthesized?

The biosynthesis of terpenoids (the primary ingredient of resin) in trees is dependent on the synthesis of isoprenoid (C-5) and its conversion into prenyl diphosphate (to form either C-10, C-15 or C-20). These compounds have been regarded as major

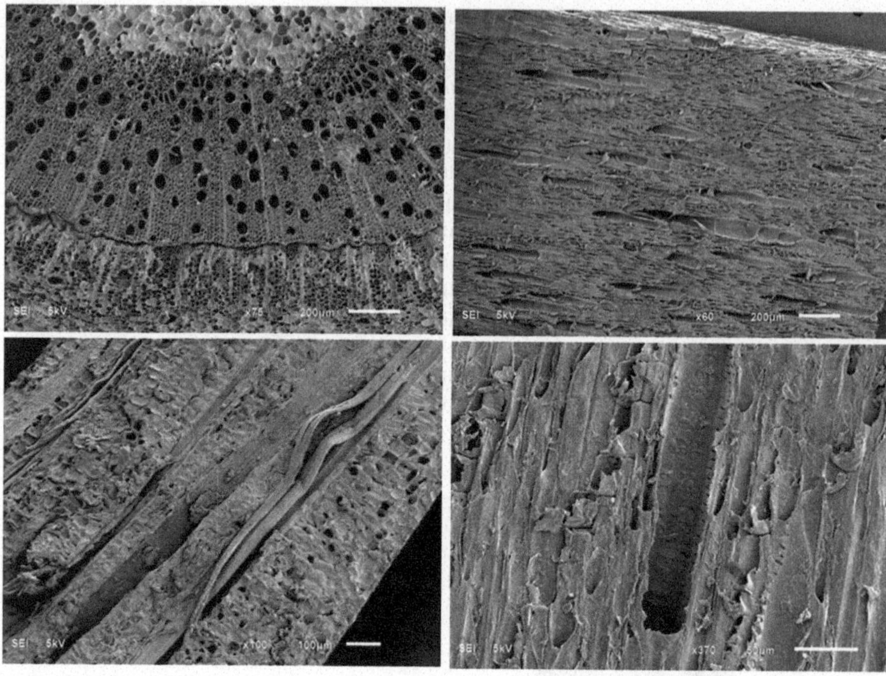

Fig. 3.3 Tree anatomy and resin production picture

substrates for producing *terpene synthases* (*TPS*) (Keeling & Bohlmann, 2006). It is understood that in some trees, *cytochrome P450 monooxygenases* activate diterpene synthase to form diterpenoid resin acids. Dimethylallyl diphosphate (DMADP) and isopentenyl diphosphate (IPP) are the known precursors of most of the isoprenoid metabolites (Lichtenthaler, Schwender, Disch, & Rohmer, 1997). These compounds are synthesized by two pathways in the plant cell (Fig. 3.4). It was observed that the IPP is formed by the reaction of acetyl CoA in the mevalonate (MVA) pathway, which is later isomerized to DMADP by isopentenyl diphosphate isomerase (IPPI) (Vogel, Wildung, Vogel, & Croteau, 1996). Mevalonate-dependent pathways and the methylerythritol 4-phosphate (MEP) pathway are present in the eukaryotes, such as bacteria and plants. In plants, the MEP pathway is responsible for the synthesis of various chemical constituents (Trapp & Croteau, 2001). The MEP pathway biosynthesizes major precursors (IPP and DMAPP) for the production of physiologically essential metabolites such as carotenoids, monoterpenes, diterpenes, isoprene, etc., and phytohormones such as abscisic acid and gibberellins and the side chains of photosynthesis-related compounds, such as plastoquinone, tocopherols, chlorophyll and phylloquinones (Trapp & Croteau, 2001). The MEP pathway produces secondary metabolites such as sterols, brassinosteroids, sesquiterpenes, triterpenes, polyterpenes, polyprenols, di-choi and prenyl moieties. It has been suggested that the MEP pathway is responsible for the synthesis of mono- and diterpenoids (Keeling & Bohlmann, 2006; Lange & Ghassemian, 2003), which make up a large part of oleoresin.

In looking at the MEP pathway, 1-deoxyxylulose 5-phosphate (DXP) synthase (DXS) is the first step in the synthesis of essential metabolites (Rodrıguez-Concepción & Boronat, 2002). This enzyme can play a rate-limiting role in the production of MEP-derived isoprenoids (Walter, Hans, & Strack, 2002). The present literature suggests that there are two types of DXS enzymes. Normally, the type I enzyme is expressed constitutively in photosynthesis-related tissues, where it is presumably involved in synthesizing isoprenoids for primary metabolism, such as the synthesis of carotenoids and phytols. In type II DXS enzymes, it is involved in the synthesis of an isoprenoid specifically for secondary metabolism (Walter et al., 2002). Previously, these two enzymes were cloned and characterized from Norway spruce trees, where both of them were homodimeric (Schmidt & Gershenzon, 2008).

The basic terpene skeleton is synthesized from geranyl pyrophosphate (GPP), farnesyl diphosphate (FPP) or geranylgeranyl diphosphate (GGPP) through the catalysis of terpene synthase (TPS) produced in monoterpenes (10-C), sesquiterpenes (15-C) and diterpenes (20-C), respectively (Davis & Croteau, 2000). The enzyme functions via divalent metal ions assist in the production of carbocation intermediates through prenyl diphosphate precursors, and they generate a wide array of cyclic and acyclic parent skeleton terpenoids. Many terpene synthases yield only one or a few closely related products, whereas some terpene synthases form complex product mixtures (Steele, Katoh, Bohlmann, & Croteau, 1998). Most of

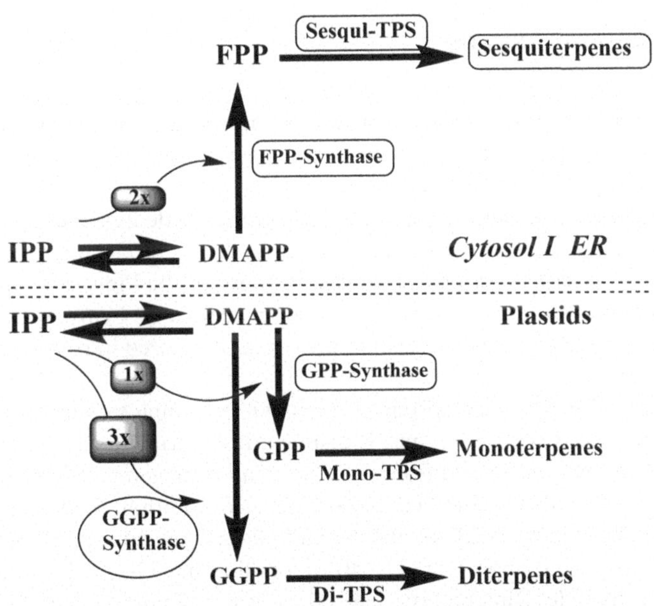

Fig. 3.4 Terpenoid biosynthesis pathways in conifers/angiosperms. *IPP* isopentenyl diphosphate, *DMAPP* dimethylallyl pyrophosphate, *PTs* prenyltransferases, *GPP* geranyl pyrophosphate, *FPP* farnesyl pyrophosphate, *GGPP* geranylgeranyl pyrophosphate, and *TPS* terpene synthases

the research on the biosynthesis of conifer terpenoids to date has focused on the terpene synthase enzymes and their genetic regulation in grand fir (Ro & Bohlmann, 2006). A study has shown that the wound-induced resin synthesis responses in the stem part of the tree are dependent on the up-regulation of terpene synthases. However, little is understood about the biochemical mechanisms of stress-induced terpenoid biosynthesis in conifers, angiosperms and *Boswellia* species (D. Martin, Tholl, Gershenzon, & Bohlmann, 2002); Fig. 3.4).

There are more than 30 genes that have been identified and isolated for the encoding of terpene synthases or cyclases. These genes form the primary skeleton of the terpenoid biosystems. A genetic comparison has been made by Mitchel-Olds et al., 1998 (Mitchell-Olds & Pedersen, 1998; Trapp & Croteau, 2001) suggesting a common evolutionary perspective at its origin. However, there is an assumption that there is a bifurcation in gymnosperm and angiosperms (Tholl, 2006).

The oxidation of oleoresin diterpenoids closely resembles the synthesis of ent-kaurenoic acid with gibberellin synthesis (Hamberger & Bohlmann, 2006; MacMillan, 2001; Ro & Bohlmann, 2006). This finding shows that diterpenic acid biosynthesis and the biosynthesis of gibberellins share a common origin. This result can also be correlated to the common nature of diTPS formation for oleoresin as well (D. E. Hall et al., 2012; S. Martin, Giannone, Andriantsitohaina, & Carmen Martinez, 2003; Zulak & Bohlmann, 2010). Hence, a correlation can also be made for *Boswellia*, which is not a known hypothesis (see Chap. 10 for more detail).

Resin-Producing Plants

In the case of conifers, Pinaceae, Araucariaceae, Cupressaceae s.l. and Podocarpaceae unequivocally produce resin in secretory structures, although resin synthesis can be induced by injury without a secretory structure in some cases. Similarly, members of the Cephalotaxaceae have resin ducts, but there is little evidence that they produce resin now, and *Sciadopitys* has not been recorded as producing resin. Taxaceae members produce terpene synthases similar to those of the Pinaceae, but they have no resin-secreting structure. Some Pinaceae genera, e.g. *Cedrus*, *Pseudotsuga*, *Pseudolarix* and *Tsuga*, produce resin in the trunk of the tree in response to trauma (Wu & Hu, 1997).

In Araucariaceae, *Araucaria* produces resin with a complex composition that can sometimes be mixed with gum, and it displays less variability in its constituents across different species. In addition, *Agathis* is another genus that produces both timber and resin with commercial importance. In addition, Podocarpaceae and Cupressaceae have also been known to include species that produce resins. In angiosperms, there are approximately 40 orders that have been shown to produce resins. These include Dipterocarpaceae, *Dracaena*, *Xanthorrhoea*, *Liquidambar* and *Altingia*, *Acacia*, *Canarium*, *Bursera* (100 species), *Boswellia* (19 species) and *Commiphora* (190 species), and they have widely been known to produce high-quality resin with potential benefits to human health.

Based on their major chemical constituents, resins are classified as either terpenoid resins or phenolic resins (Langenheim, 2003). The terpenoid resins are the largest diverse class of plant compounds with over 30,000 known terpenoid compounds. These resins occur in most conifer families, and some authors refer to them as conifer resins. The leading and highly exploited example of these resins is extracted from pine trees (*Pinus*), with global annual production of over 300,000 tons (McGarvey & Croteau, 1995). This resin is primarily used as a solvent for thinning oil-based paints or as varnish that provides layers of protection for wooden furniture or is used as adhesives or as a novel source of organic compounds (Stenius, 2000). It is worth noting that the fossilized resin known as amber, which has been used in jewellery production since the Stone Age (approximately 13,000 years ago), is of coniferous origin (Grimaldi, 2009; Wolfe et al., 2009). Interestingly, the Maya and Aztec people have also adopted incense by burning the resin of the coniferous copal tree (Mckenna & Hughes, 2014).

Composition of Frankincense Resin

Frankincense is primarily composed of a mucus-like cluster (12–23%), essential oil (5–15%) and lipophilic part (55–66%). This composition has shown quite a bit of variation depending on the species and grades of resin (Halliwell & Gutteridge, 2015) (see Chap. 10 for more detail). Frankincense contains volatile compounds, with more than 20 monoterpenes and 28 sesquiterpenes. The lipophilic parts are composed of terpenoids, including the boswellic acids (BAs), which are the chemotaxonomic markers of *Boswellia* (Miyamoto, Martinez, Medeiros, & Di Mascio, 2014; Pospíšil, Prasad, & Rác, 2014). Recently, Al-Harrasi et al. (2018) elucidated the biosynthesis of boswellic acid and its derivatives in light of the identification of new precursors from its biosynthesis in *Boswellia sacra* (Al-Harrasi et al., 2018). Certainly, comprehensive physiological investigations about frankincense trees are mandatory to elucidate the exact order of steps for the biosynthesis of AKBA (Fig. 3.3; see Chap. 10 for more information).

Resin Production After Tapping

The quality and quantity of frankincense is dependent on the (i) age, (ii) stem diameter, (iii) number of previous and current harvests and (iv) environmental conditions (Langenheim, 2003). Moreover, studies have shown that the resin content, composition, colour, grade and quantity also depend greatly on the above four major factors. Additional factors such as the overall population structure, elevation, drought or more water, nutrient run-off during rainfalls, insect attacks and/or camel grazing (Caro, 2005; Steenkamp, Vogel, Fuls, Van Rooyen, & Van Rooyen, 2008) and elevation (Groenendijk, Eshete, Sterck, Zuidema, & Bongers, 2012; Mengistu, Sterck, Fetene, & Bongers, 2013) contribute to the overall physical and chemical properties of frankincense.

There are various tapping methods for extracting resin from *Boswellia*. The traditional methods include using a blunt end knife or sickle or an iron scraper called the *Mengaf*. Since the axial and radial canals are located at the parenchymal region immediately after the outer bark, any tapping will help the resin to ooze out (Gebrehiwot, Muys, Haile, & Mitloehner, 2003; Mengistu et al., 2013; Tolera, Sass-Klaassen, Eshete, Bongers, & Sterck, 2013) (Fig. 3.5). However, the deeper the cut is, the lower the quantity of resin oozing out, and deep cuts can lead to an irreparable loss to tree growth and development. The white milky exudate that comes out of the tree is air-dried to ensure that it hardens into globular pears or tears (Gebrehiwot et al., 2003).

Fig. 3.5 Resin oozing out of axillary bud and tree stems of *B. sacra* in Dhofar, Oman

For *B. papyrifera* and *B. sacra*, the tapping is 1–1.5 mm deep with an area of 2–2.5 cm^2. For a *Boswellia* tree with a diameter of approximately 15–35 cm, the incisions should be made in opposite directions at 2 feet high. The number of cuts should not exceed 6–8 cuts per tree. In Oman, the cuts are traditionally made to face the sun to enable the resin to dry quickly. This practice can be performed six to ten times during the entire season in parts of Ethiopia, Sudan and Yemen. Usually, the initial two to three tappings are performed within 21 days. However, the duration is shortened during the later tappings. Sometimes, the resin is collected earlier to avoid moisture exposure because increasing this exposure to moisture can change the colour and shape of the frankincense tears (Gebrehiwot et al., 2003). Similarly, significant variations are expected to exist, e.g. between different species, provenances, climates, resin flow-inducing treatments, age classes and population structures (Roberts, 2003; Rodriguez & Redman, 2008). Most studies on resin flow have been conducted under controlled conditions, when opening small wounds for a short period (Gaylord et al., 2007; Lombardero, Ayres, Lorio Jr., & Ruel, 2000; Novick, Katul, McCarthy, & Oren, 2012) in saplings or young trees (Knebel, Robison, Wentworth, & Klepzig, 2008; Ruel et al., 1998).

Uses of Frankincense

Frankincense has been widely used for its medicinal, cultural and economic benefits. It is categorized as a fresh balsamic resin, thus making it a valuable source in the development of a wide variety of perfumes (see Chaps. 1 and 10 for more information). There have been numerous studies on the uses of frankincense. The chemistry and biological activities of boswellic acids and other terpenoids in the *Boswellia* genus have recently been covered by Al-Harrasi et al. (Al-Harrasi et al., 2018). Moreover, the application of frankincense in modern medicine has been documented by Ammon in his recently published book (Ammon, 2006).

At the industrial level, frankincense-based products have been extensively developed ((Stiles, 2003); Ethiopian Forestry Action Program (EFAP), 1993; Food and Agriculture Organization (FAO), 1995; (Index, 1996)). Various pharmaceutical products have been produced as shown in the work of Brown (1995), Michie and Cooper (1991), Rani et al. (1996) and Wichtl and Bisset (1994). In addition, frankincense-related products have also been used in the food industry (R. L. Hall & Oser, 1965; Tucker, 1986) and in flavour, liqueur and beverage (Rani et al., 1996) industries. More recently, many cosmetic and perfumery industries have also helped to promote its uses (Farah, 1994; Tucker, 1986).

Production of Frankincense

For over 5000 years, people have treasured, used and traded frankincense. The recent modern wholesale price offered for internationally exported frankincense is based on its type and grade, which amounts to US$ 3000 to 6000 per ton (Table 3.2). For example, two types are distinguished for Somalian frankincense, with 'Maidi' being the best, which is normally sold at the higher end of the price range, and 'Bayo', with annual exports of 800 and 200 tons, respectively. The Indian type amounts to 200 tons annually, while all other African sources are referred to as the Eritrean type, representing the highest share of the market at approximately 2000 tons annually as of the 1987 published data, but it is priced at the lower end of this range (34). No data are available on the current international export market share of the different types produced by *B. sacra* from Yemen and Oman. Likewise, no annual production data have been recorded. This information should receive immediate attention for long-term sustainability and ecosystem management.

Table 3.2 Grading of resin obtained from *Boswellia* and other species

Grades	Size (mm)	Colour
1A	6 mm	White
IB	6	Creamy white
2	>4 > 6	White
3	>2 to >4	White
4 – special	Any size	Brown
4 – normal	Any size	Black
5	Powder and bark	

Source: CIFOR 2011

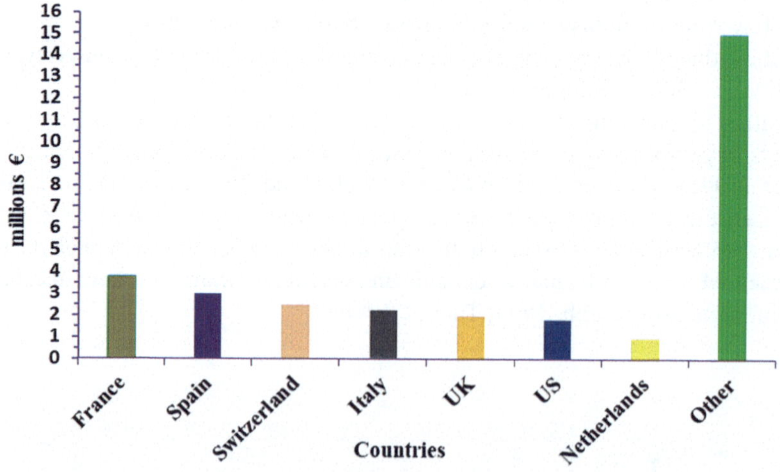

Fig. 3.6 Export of resins from Germany to various destinations (Eurostat, 2015)

Oleoresin gums, which are derived from frankincense trees, are important sources of income for sustaining economies at regional and national levels, and they are a source of quick cash for indigenous peoples. The extraction of frankincense has often been categorized as a non-timber forest product by the FAO (1995). In some studies from the past decade, the frankincense harvest contributes approximately 35% of the domestic income of a rural family (Lemenih, Abebe, & Olsson, 2003). A tapper can collect approximately 1000–1500 kg of incense per annum and will receive a net income of US$ 100–150 (Aregawi, 1997). Women accrue an annual average income of US$ 160 (Gebremedhin, 1997), assuming 10 months of off-farm employment.

In addition to local economies, the frankincense tree and its products are currently in high demand in international markets for the following uses:

- Essential oil extraction and perfumery industries
- Active ingredients in herbal recipes in therapeutic medicines
- Domestic use in homes and for celebrating occasions
- Beekeeping and livestock grazing and fodder
- Wood products
- Research and development industry
- Religious activities
- Tanning industries
- Cosmetics industries
- Food industries, flavouring in bakeries
- Fabrication of varnishes, adhesives, fumigation powders, dye manufacturing

The major producer of frankincense is Africa, followed by India. The major exporters are Germany, France and North America. Moreover, due to its international demand, this compound plays versatile roles in the foreign exchange of Africa (Farah, 1994; Gebremedhin, 1997; Thulin & Claeson, 1991; Tucker, 1986). According to the Natural Gum Processing and Marketing Enterprise (NGPME) of Ethiopia, in 1988, the export of natural gum was only 663.4 tons, and at a domestic level, 1084.4 tons were traded, which cumulatively accounted for revenue of US$ 1,296,008 (Gebremedhin, 1997). The above-mentioned data do not include unrecorded data, which would include highly local trading and transborder trade with neighbouring countries such as Kenya and Somalia. Even today, Ethiopia and Somalia produce and export a major portion of the myrrh and frankincense to the world (Coulter, 1987; Farah, 1994; Wahab et al., 1987). Due to the increasing demand in Germany, resin imports were increased by 6.6% over consecutive years from 2010 to 2014, which accounted for an average of 55 million euros. This increase was primarily due to the cost of resin, although the imports were the same (13,000 tons) (Fig. 3.6).

The aromatic resin that is generally obtained from the genera *Boswellia* and *Commiphora* is a source of income for poor households (Lemenih, Wiersum, Woldeamanuel, & Bongers, 2014). The two most important aromatic resins are frankincense, which is obtained from different *Boswellia* species, such as *B. frereana*, *B. neglecta*, *B. papyrifera* and *B. sacra*, and myrrh, which is obtained from

Commiphora species such as *C. myrrha*, *C. hadi* and *C. holtziana*, either naturally or after tapping (Farah, 1994; Hassan et al., 2011; Tadesse, Desalegn, & Alia, 2007). According to its brochure, it exports 1000–1500 tons of natural gum and resin, which makes up 40% of the country's total exports. Essential oil production usually amounts to 200–250 euros/kg for frankincense and approximately 300 euros/kg for myrrh, while the global prices of raw myrrh and frankincense are usually 2–3 euros/kg. The global market demand for this aromatic resin is 6000 metric tons per annum, for which the major exporters are Ethiopia (3000 t/a), Eritrea (400 t/a), Kenya (1500 t/a) and Somalia (1200 t/a) (Chikamai & Casadei, 2005; Ogbazghi, Bongers, Rijkers, & Wessel, 2006). The major importers of aromatic resin are China, Europe, the Middle East, Japan, the USA and North Africa, while the emerging markets are South America and Eastern Europe. The exports from the Horn of Africa are still considered lower than what they could be (Lemenih & Kassa, 2011). However, most of this resin is often contaminated, which does not help rural communities to earn economic benefits. Contaminants such as sand, small stones, dirt and larger bark pieces prevent producers from obtaining higher prices.

Marketing of Frankincense

There are five major steps in the marketing of frankincense, namely, (i) preharvest, (ii) harvest, (iii) postharvest, (iv) sales and distribution at the national market and (v) cross-country and cross-continent channels of supply. The price of the raw produce and final product could vary from 200 to 500%; thus, the profit distribution at the grassroots level is meagre. Figure 3.7 shows the market chain of the frankincense supply until the final product, where the last industrial segment profit increases by fivefold, especially in the case of cosmetics.

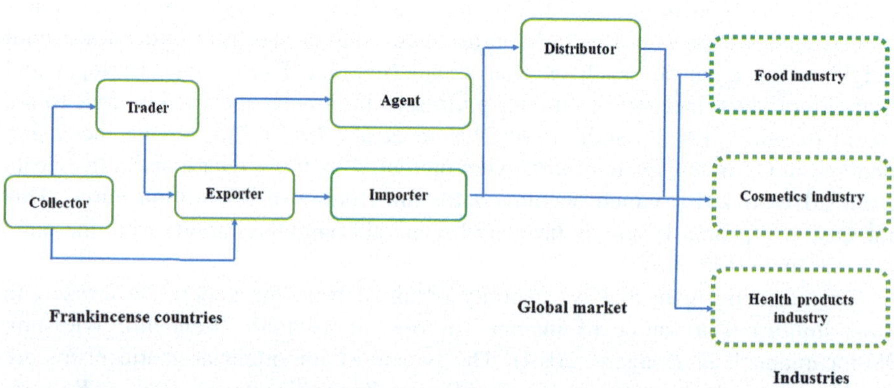

Fig. 3.7 Resin market chain from produce to industrial produce (Eurostate, 2015)

Climatic Changes Influencing Frankincense Production

The resin industry of Southern Europe demands more efficient tapping methods and forest management plans to increase the supply and enhance competitiveness in a global market (Cunningham, Brinckmann, Kulloli, & Schippmann, 2018). Understanding the intra- and interannual effects of climate conditions and soil water availability on resin yields in the pine stands used for tapping is essential for improving techniques and plans in the resin industry. This approach becomes particularly important under the current climate change scenario, in which projections for the Mediterranean region indicate more frequent extreme climate conditions, which could alter water availability and impact the delivery of ecosystem services (Groenendijk et al., 2012; Mengistu et al., 2013; Rehman et al., 2018; Tolera et al., 2013). Climate projections for the Mediterranean region predict a pronounced decrease in precipitation (especially in the warm season), more extended droughts, marked warming reaching a maximum in summer, the concentration of rainfall during shorter periods of the year and a greater occurrence of extremely high temperature events (Giorgi & Lionello, 2008). The negative effects of the already ongoing climate change on non-wood forest products, such as Mediterranean pine nuts, have been described for Central Spain, in the middle basin of the Douro River (Mutke & Barthlott, 2005). Climate change may also have unpredictable effects on resin production, although research has shown that climate conditions can affect resin flow in two primary ways: directly, because the temperature changes the resin fluidity (Al-Qurashi & Awad, 2011), and indirectly, because climate factors, such as temperature and water stress, have a major effect on the tree anatomical structure and the physiological processes involved in resin biosynthesis and secretion (Rodríguez-García et al., 2018). Because of the positive relationships between resin flow (as plant secondary chemistry) and pest resistance (e.g. Kolosova & Bohlmann, 2012; Mumm & Hilker, 2006), the described climate effects on the resin flow could be directly applicable to the tree's defence capacity against bark-boring insects and fungal pathogens.

References

Al-Harrasi, A., Rehman, N. U., Khan, A. L., Al-Broumi, M., Al-Amri, I., Hussain, J., … Csuk, R. (2018). Chemical, molecular and structural studies of Boswellia species: β-Boswellic Aldehyde and 3-epi-11β-Dihydroxy BA as precursors in biosynthesis of boswellic acids. *PLoS One, 13*(6), e0198666.

Al-Qurashi, A. D., & Awad, M. A. (2011). 5-Aminolevulinc acid increases tree yield and improves fruit quality of 'Rabia' and 'Sukkariat-Yanbo' date palm cultivars under hot arid climate. *Scientia Horticulturae, 129*(3), 441–448. https://doi.org/10.1016/j.scienta.2011.04.014

Ammon, H. (2006). Boswellic acids in chronic inflammatory diseases. *Planta Medica, 72*(12), 1100–1116.

Aregawi, B. (1997). *Preliminary survey on forest products utilization and marketing in Tigray*. Mekelle, Ethiopia: Tigray Bureau of Agricultural Development and Natural Resources.

Boschiero, F. A., & Tomazzello-Filho, M. (2012). Anatomical aspects of resin canals and oleoresin production in pine trees. In *Resin: Biology, chemistry and applications* (pp. 67–86). Kerala, India: Research Signpost.

Brown, J. H. (1995). *Macroecology*. University of Chicago Press.

Caro, T. (2005). *Antipredator defenses in birds and mammals*. University of Chicago Press.

Chikamai, B., & Casadei, E. (2005). *Production and marketing of gum resins*. Frankincense, Myrrh and Opoponax.

Coulter, J. (1987). *Market study for frankincense and myrrh from Somalia*. Chatham, UK: National Resources Institute.

Cunningham, A., Brinckmann, J., Kulloli, R., & Schippmann, U. (2018). Rising trade, declining stocks: The global gugul (Commiphora wightii) trade. *Journal of Ethnopharmacology, 223*, 22.

Davis, E. M., & Croteau, R. (2000). *Cyclization enzymes in the biosynthesis of monoterpenes, sesquiterpenes, and diterpenes Biosynthesis* (pp. 53–95). Springer.

Eslamieh, J. (2010). Creating "perfect" Boswellia. *Cactus and Succulent Journal, 82*(3), 126–131.

Farah, A. Y. (1994). *The milk of the Boswellia forests: Frankincense production among the pastoral Somali*. EPOS, Environmental Policy and Society.

Gaylord, T. G., Barrows, F. T., Teague, A. M., Johansen, K. A., Overturf, K. E., & Shepherd, B. (2007). Supplementation of taurine and methionine to all-plant protein diets for rainbow trout (*Oncorhynchus mykiss*). *Aquaculture, 269*(1–4), 514–524.

Gebrehiwot, K., Muys, B., Haile, M., & Mitloehner, R. (2003). Introducing *Boswellia papyrifera* (Del.) Hochst and its non-timber forest product, frankincense. *International Forestry Review, 5*(4), 348–353.

Gebremedhin, T. (1997). *Boswellia papyrifera* (Del.) hochst. From Western Tigray: Opportunities, constraints and seed germination responses.

Giorgi, F., & Lionello, P. (2008). Climate change projections for the Mediterranean region. *Global and Planetary Change, 63*(2–3), 90–104.

Grimaldi, D. (2009). Pushing back amber production. *Science, 326*(5949), 51–52.

Groenendijk, P., Eshete, A., Sterck, F. J., Zuidema, P. A., & Bongers, F. (2012). Limitations to sustainable frankincense production: Blocked regeneration, high adult mortality and declining populations. *Journal of Applied Ecology, 49*(1), 164–173.

Hall, D. E., Zerbe, P., Jancsik, S., Quesada, A. L., Dullat, H., Madilao, L. L., ... Bohlmann, J. (2012). Evolution of conifer diterpene synthases: Diterpene resin acid biosynthesis in lodgepole pine and jack pine involves monofunctional and bifunctional diterpene synthases. *Plant Physiology*, 112.208546.

Hall, R. L., & Oser, B. L. (1965). Recent progress in consideration of flavoring ingredients under food additives amendment. 3. Gras substances. *Food Technology, 19*(2 P 2), 151.

Halliwell, B., & Gutteridge, J. M. (2015). *Free radicals in biology and medicine*. USA: Oxford University Press.

Hamberger, B., & Bohlmann, J. (2006). *Cytochrome P450 mono-oxygenases in conifer genomes: Discovery of members of the terpenoid oxygenase superfamily in spruce and pine*. Portland Press Limited.

Hassan, B., Glover, E. K., Luukkanen, O., Chikamai, B., Jamnadass, R., Iiyama, M., & Kanninen, M. (2011). The role of Boswellia and Commiphora species in rural livelihood security and climate change adaptation in the Horn of Africa: Case study of Northeastern Kenya. *International Journal of Social Forestry, 4*(1), 86–112.

Index, M. (1996). *An encyclopedia of chemicals, drugs and biologicals* (12th ed., p. 8697). USA: Merck and Co. INC.

Keeling, C. I., & Bohlmann, J. (2006). Diterpene resin acids in conifers. *Phytochemistry, 67*(22), 2415–2423.

Knebel, L., Robison, D. J., Wentworth, T. R., & Klepzig, K. D. (2008). Resin flow responses to fertilization, wounding and fungal inoculation in loblolly pine (*Pinus taeda*) in North Carolina. *Tree physiology, 28*(6), 847–853.

Kolosova, N., & Bohlmann, J. (2012). *Conifer defense against insects and fungal pathogens Growth and defence in plants* (pp. 85–109). Springer.

Krokene, P., & Nagy, N. E. (2012). Anatomical aspects of resin-based defences in pine. *Pine Resin: Biology, Chemistry and Applications*, 67–86.

Lange, B. M., & Ghassemian, M. (2003). Genome organization in *Arabidopsis thaliana*: a survey for genes involved in isoprenoid and chlorophyll metabolism. *Plant Mol Biol, 51*(6), 925–948.

Langenheim, J. H. (2003). *Plant resins: chemistry, evolution, ecology, and ethnobotany*. Oregon, USA: Timber Press.

Lemenih, M., Abebe, T., & Olsson, M. (2003). Gum and resin resources from some Acacia, Boswellia and Commiphora species and their economic contributions in Liban, south-east Ethiopia. *Journal of Arid Environments, 55*(3), 465–482.

Lemenih, M., & Kassa, H. (2011). *Opportunities and challenges for sustainable production and marketing of gums and resins in Ethiopia*. Bogor, Indonesia: Cifor.

Lemenih, M., Wiersum, K., Woldeamanuel, T., & Bongers, F. (2014). Diversity and dynamics of management of gum and resin resources in Ethiopia: a trade-off between domestication and degradation. *Land degradation & development, 25*(2), 130–142.

Lichtenthaler, H. K., Schwender, J., Disch, A., & Rohmer, M. (1997). Biosynthesis of isoprenoids in higher plant chloroplasts proceeds via a mevalonate-independent pathway. *FEBS letters, 400*(3), 271–274.

Lombardero, M., Ayres, M. P., Lorio, P. L., Jr., & Ruel, J. J. (2000). Environmental effects on constitutive and inducible resin defences of *Pinus taeda*. *Ecology letters, 3*(4), 329–339.

MacMillan, J. (2001). Occurrence of gibberellins in vascular plants, fungi, and bacteria. *Journal of plant growth regulation, 20*(4), 387–442.

Martín-Ramos, P., Fernández-Coppel, I. A., Ruíz-Potosme, N. M., & Martín-Gil, J. (2018). Potential of ATR-FTIR Spectroscopy for the Classification of Natural Resins. *Biology, Engineering, Medicine and Science Reports, 4*(1).

Martin, D., Tholl, D., Gershenzon, J., & Bohlmann, J. (2002). Methyl jasmonate induces traumatic resin ducts, terpenoid resin biosynthesis, and terpenoid accumulation in developing xylem of Norway spruce stems. *Plant physiology, 129*(3), 1003–1018.

Martin, S., Giannone, G., Andriantsitohaina, R., & Carmen Martinez, M. (2003). Delphinidin, an active compound of red wine, inhibits endothelial cell apoptosis via nitric oxide pathway and regulation of calcium homeostasis. *British journal of pharmacology, 139*(6), 1095–1102.

McGarvey, D. J., & Croteau, R. (1995). Terpenoid metabolism. *The Plant Cell, 7*(7), 1015.

Mckenna, D. J., & Hughes, K. (2014). *The incense bible: Plant scents that transcend world culture, medicine, and spirituality*. Routledge.

Mengistu, T., Sterck, F. J., Fetene, M., & Bongers, F. (2013). Frankincense tapping reduces the carbohydrate storage of Boswellia trees. *Tree physiology, 33*(6), 601–608.

Michie, C. A., & Cooper, E. (1991). Frankincense and myrrh as remedies in children. *Journal of the Royal Society of Medicine, 84*(10), 602.

Mitchell-Olds, T., & Pedersen, D. (1998). The molecular basis of quantitative genetic variation in central and secondary metabolism in Arabidopsis. *Genetics, 149*(2), 739–747.

Miyamoto, S., Martinez, G. R., Medeiros, M. H., & Di Mascio, P. (2014). Singlet molecular oxygen generated by biological hydroperoxides. *Journal of Photochemistry and Photobiology B: Biology, 139*, 24–33.

Mumm, R., & Hilker, M. (2006). Direct and indirect chemical defence of pine against folivorous insects. *Trends in plant science, 11*(7), 351–358.

Mutke, J., & Barthlott, W. (2005). Patterns of vascular plant diversity at continental to global scales. *Biologiske skrifter, 55*(4), 521–531.

Novick, K., Katul, G., McCarthy, H., & Oren, R. (2012). Increased resin flow in mature pine trees growing under elevated CO2 and moderate soil fertility. *Tree physiology, 32*(6), 752–763.

Ogbazghi, W., Bongers, F., Rijkers, T., & Wessel, M. (2006). Population structure and morphology of the frankincense tree *Boswellia papyrifera* along an altitude gradient in Eritrea. *Journal of the Drylands, 1*(1), 85–94.

Ogbazghi, W., Rijkers, T., Wessel, M., & Bongers, F. (2006). Distribution of the frankincense tree *Boswellia papyrifera* in Eritrea: the role of environment and land use. *Journal of Biogeography, 33*(3), 524–535.

Phillips, M. A., & Croteau, R. B. (1999). Resin-based defenses in conifers. *Trends in plant science, 4*(5), 184–190.

Pospíšil, P., Prasad, A., & Rác, M. (2014). Role of reactive oxygen species in ultra-weak photon emission in biological systems. *Journal of Photochemistry and Photobiology B: Biology, 139*, 11–23.

Rani, M. S., Foster, G. R., Leung, S., Leaman, D., Stark, G. R., & Ransohoff, R. M. (1996). Characterization of β-R1, a gene that is selectively induced by interferon β (IFN-β) compared with IFN-α. *Journal of Biological Chemistry, 271*(37), 22,878–22,884.

Rehman, N. U., Ali, L., Al-Harrasi, A., Mabood, F., Al-Broumi, M., Khan, A. L., … Csuk, R. (2018). Quantification of AKBA in *Boswellia sacra* Using NIRS Coupled with PLSR as an Alternative Method and Cross-Validation by HPLC. *Phytochemical Analysis, 29*(2), 137–143.

Ro, D.-K., & Bohlmann, J. (2006). Diterpene resin acid biosynthesis in loblolly pine (*Pinus taeda*): functional characterization of abietadiene/levopimaradiene synthase (PtTPS-LAS) cDNA and subcellular targeting of PtTPS-LAS and abietadienol/abietadienal oxidase (PtAO, CYP720B1). *Phytochemistry, 67*(15), 1572–1578.

Roberts, M. R. (2003). 14–3-3 proteins find new partners in plant cell signalling. *Trends in plant science, 8*(5), 218–223.

Rodrıguez-Concepción, M., & Boronat, A. (2002). Elucidation of the methylerythritol phosphate pathway for isoprenoid biosynthesis in bacteria and plastids. A metabolic milestone achieved through genomics. *Plant physiology, 130*(3), 1079–1089.

Rodríguez-García, A., Madrigal, J., González-Sancho, D., Gil, L., Guijarro, M., & Hernando, C. (2018). Can prescribed burning improve resin yield in a tapped *Pinus pinaster* stand? *Industrial Crops and Products, 124*, 91–98.

Rodriguez, R., & Redman, R. (2008). More than 400 million years of evolution and some plants still can't make it on their own: plant stress tolerance via fungal symbiosis. *Journal of experimental botany, 59*(5), 1109–1114.

Ruel, J. J., Ayres, M. P., Lorio, J., & Peter, L. (1998). Loblolly pine responds to mechanical wounding with increased resin flow. *Canadian Journal of Forest Research, 28*(4), 596–602.

Schmidt, A., & Gershenzon, J. (2008). Cloning and characterization of two different types of geranyl diphosphate synthases from Norway spruce (*Picea abies*). *Phytochemistry, 69*(1), 49–57.

Steele, C. L., Katoh, S., Bohlmann, J., & Croteau, R. (1998). Regulation of oleoresinosis in grand fir (*Abies grandis*): differential transcriptional control of monoterpene, sesquiterpene, and diterpene synthase genes in response to wounding. *Plant physiology, 116*(4), 1497–1504.

Steenkamp, C. J., Vogel, J., Fuls, A., Van Rooyen, N., & Van Rooyen, M. W. (2008). Age determination of *Acacia erioloba* trees in the Kalahari. *Journal of Arid Environments, 72*(4), 302–313.

Stenius, P., ed. (2000). Forest Products Chemistry. *Papermaking Science and Technology. 3. Finland*, 73–76.

Stiles, W. B. (2003). Qualitative research: Evaluating the process and the product. *Handbook of clinical health psychology, 24*, 477–500.

Tadesse, W., Desalegn, G., & Alia, R. (2007). Natural gum and resin bearing species of Ethiopia and their potential applications. *Forest Systems, 16*(3), 211–221.

Taniguchi, Y., Taniguchi, H., Yamada, M., Matsukura, Y., Koizumi, H., Furihata, K., & Shindo, K. (2014). Analysis of the components of hard resin in hops (*Humulus lupulus* L.) and structural elucidation of their transformation products formed during the brewing process. *Journal of Agricultural and Food Chemistry, 62*(47)., 11,602–11,612.

Tholl, D. (2006). Terpene synthases and the regulation, diversity and biological roles of terpene metabolism. *Current opinion in plant biology, 9*(3), 297–304.

Thulin, M., & Claeson, P. (1991). The botanical origin of Scented Myrrh (Blssabol or Habak Hadi). *Economic Botany, 45*(4), 487–494.

Tolera, M., Sass-Klaassen, U., Eshete, A., Bongers, F., & Sterck, F. J. (2013). Frankincense tree recruitment failed over the past half century. *Forest ecology and management, 304*, 65–72.

Trapp, S. C., & Croteau, R. B. (2001). Genomic organization of plant terpene synthases and molecular evolutionary implications. *Genetics, 158*(2), 811–832.

Tucker, G. (1986). Pharamacokinetics of local anaesthetics. *British Journal of Anaesthesia, 58*(7), 717–731.

Vogel, B. S., Wildung, M. R., Vogel, G., & Croteau, R. (1996). Abietadiene synthase from grand fir (*Abies grandis*) cDNA isolation, characterization, and bacterial expression of a bifunctional diterpene cyclase involved in resin acid biosynthesis. *Journal of Biological Chemistry, 271*(38), 23,262–23,268.

Wahab, S. A., Aboutabl, E., El-Zalabani, S., Fouad, H., De Pooter, H., & El-Fallaha, B. (1987). The essential oil of olibanum. *Planta medica, 53*(04), 382–384.

Walter, M. H., Hans, J., & Strack, D. (2002). Two distantly related genes encoding 1-deoxy-d-xylulose 5-phosphate synthases: differential regulation in shoots and apocarotenoid-accumulating mycorrhizal roots. *The Plant Journal, 31*(3), 243–254.

Wichtl, M., & Bisset, N. (1994). *Herbal Drugs and Phytopharmaceuticals, Med. Stuttgart: Pharm Scientific Publ.*.

Wilkens, R. T., Ayres, M. P., Lorio, P. L., & Hodges, J. D. (1998). Environmental effects on pine tree carbon budgets and resistance to bark beetles. In *The productivity and sustainability of southern forest ecosystems in a changing environment* (pp. 591–616). Springer.

Wolfe, A. P., Tappert, R., Muehlenbachs, K., Boudreau, M., McKellar, R. C., Basinger, J. F., & Garrett, A. (2009). A new proposal concerning the botanical origin of Baltic amber. *Proceedings of the Royal Society of London B: Biological Sciences*, rspb20090806.

Wu, H., & Hu, Z.-h. (1997). Comparative anatomy of resin ducts of the Pinaceae. *Trees, 11*(3), 135–143.

Zulak, K. G., & Bohlmann, J. (2010). Terpenoid biosynthesis and specialized vascular cells of conifer defense. *Journal of Integrative Plant Biology, 52*(1), 86–97.

Chapter 4
Frankincense Tree Physiology and Its Responses to Wounding Stress

Wounding Stress Physiology in Plants

Unlike animals, plants have been bestowed with a more complex stress-coping mechanism, and it changes according to growth dynamics and environmental conditions such as light, nutrients and moisture. Plants often have to face difficult growth and ecological circumstances, viz. drought, salinity, heat, minerals and heavy metals, creating stress and hindering growth dynamics (A. L. Khan, Hussain, Al-Harrasi, Al-Rawahi, & Lee, 2015; A. L. Khan et al., 2017). In addition, either human-induced or natural stresses such as wounding or physical injury by wind, herbivory and tapping cause plants to use energy to continuously control stress factors and cofactors (Hyodo, 2017; Jacobo-Velázquez, González-Agüero, & Cisneros-Zevallos, 2015; Koldenkova & Hatsugai, 2018; Koo, 2018; Torres-Contreras et al., 2018). Thus, the plant or the damaged part is vulnerable to pathogenic attacks (Davey, Stals, Panis, Keulemans, & Swennen, 2005; Farmer, Alméras, & Krishnamurthy, 2003; León, Rojo, & Sánchez-Serrano, 2001; Wasternack et al., 2006; Zarate, Kempema, & Walling, 2007). Wounding or physical injury creates oxidative stress by producing highly reactive oxygen species (ROS), which in turn also damage the unsaturated lipid layer and related membrane, thereby affecting both the active and passive transport systems across cell tissues and exposing the cellular system to exogenous stress (Davey et al., 2005; Hyodo, 2017; Koo, 2018; León et al., 2001). ROS include superoxide radicals (O_2^-), hydrogen peroxide (H_2O_2), hydroxyl radicals (OH^-) and single radicals ($^1O_2$) (Y.-H. Kim et al., 2015). In addition, both long- and short-distance travel physiologies coupled with the plant's anatomy are essential for obstructing stress damage. The ability of a plant to 'sense' a wound can occur through Ca^{2+}, ROS waves and/or electrical signals (Choi et al., 2017; Koldenkova & Hatsugai, 2018). An oxidative burst is followed by electrical signals in the form of ion fluxes (Oyarce & Gurovich, 2011), which also simultaneously activate chemical messengers; for example, terpenes, methyl salicylate, methyl benzoate, ethylene and especially jasmonic acid (Kovalchuk et al., 2013; R. E. McDowell, Amsler, Li,

A. Al-Harrasi et al., *Biology of Genus Boswellia*,
https://doi.org/10.1007/978-3-030-16725-7_4

Lancaster Jr, & Amsler, 2015). The evocation of priming effects triggered by either systemic acquired resistance (SAR), through biotic stresses, or systemic acquired acclimation (SAA) occurred through abiotic stimuli (Koo, 2018). These systemic signals can occur from cell-to-cell, organ-to-organ and even plant-to-plant communication systems (Holbrook & Knoblauch, 2018). The greater the stress is, the greater the lipid breakdown and its related products (Farmer et al., 2003; A. L. Khan, Waqas, et al., 2015b; I. S. Kim, Koppula, Park, & Choi, 2017). In SAR, after the activation of ROS, phytohormones such as jasmonic acid (JA) are activated by lipid-derived signals from α-linolenic acid (Wasternack et al., 2006; Zarate et al., 2007). This signalling starts with a 13-lipoxygenase (13-LOX) catalysing the oxygenation of α-linolenic acid followed by the metabolism of the LOX product 13-hydroperoxy linolenic acid by an allene oxide synthase (AOS). Subsequently, 12-oxo-phytodienoic acid (PDA) is formed by allene oxide cyclase (AOC). Finally, a reductase and three β-oxidation steps form JA (Savatin, Gramegna, Modesti, & Cervone, 2014). During wounding stress, lipid peroxidation and JA synthesis increase markedly. In addition to JA, salicylic acid (SA) has also been perceived as a major biotic stress respondent, especially to herbivory or pathogenic infection after wounding or physical injury (Y.-B. Kim et al., 2009). SA can regulate various physiological and biochemical responses, including the seed germination (Y.-H. Kim et al., 2015), photosynthesis pigment content (Fariduddin, Hayat, & Ahmad, 2003), transpiration rate, stomatal conductance, internal CO_2 concentration (Sh Hayat, Yadav, Ali, & Ahmad, 2010), nitrate reductase activity (S Hayat, Fariduddin, Ali, & Ahmad, 2005), regulation of endogenous hormones (Shakirova, Sakhabutdinova, Bezrukova, Fatkhutdinova, & Fatkhutdinova, 2003) and modulation of antioxidant activities (Zhou & Underhill, 2015). In addition to SA, ethylene has also been one of the major players in wounding stress responses (Klay et al., 2018). It responds to mechanical wounding, water deficit, flooding, low temperatures and irradiation (Hyodo, 2017). In turn, it cues the secondary responses of the plant apparatus against particular stress conditions. Secondary responses are activated through the SAR-dependent epigenetic mechanisms encountered during the RNA-directed methylation of DNA, coupling the accumulation of the up- or down-regulation for various genetic determinants of phenotypic and morphological responses. These systemic signals include cell-to-cell and organ-to-organ (shoot-to-shoot, root-to-root, root-to-shoot and shoot-to-root) interactions and possibly plant-to-plant communication.

Wounding or Tapping of Frankincense Trees

This section concerns the tapping or wounding of *Boswellia* trees with an emphasis on economically important species such as *B. sacra*, *B. papyrifera*, *B. carterii* and *B. serrata*. This tapping is primarily employed for harvesting resins and is quite common in Ethiopia, Sudan and Yemen. The tree is cut with the help of an axe

Fig. 4.1 Wounding stress in the stem of *B. sacra*

during the dry season over a 2–3 cm area and 1–2 mm depth (Tadesse, Feleke, & Eshete, 2004; Fig. 4.1). This tapping can be performed four to eight times depending on the tree size, age and location (Mengistu, Sterck, Anten, & Bongers, 2012). After wounding, the tree responds by oozing a milk-like viscous liquid (Mengistu, Sterck, Fetene, & Bongers, 2013). This liquid gradually spreads over the wounded part, and an additional amount is continuously released over time depending on the tree health and age (Shahzad et al., 2016; Fig. 4.1). Strong wind and heat within the environmental continuum evaporates the water from the liquid, leaving a crystal-like material that is hardened in 4–8 days (Rijkers, Ogbazghi, Wessel, & Bongers, 2006). These clumps are then traded or used domestically (Farah, 1994). The resulting resin is composed of hundreds of mono-, di- and triterpenoids, all of which contribute to the overall activity of frankincense (Al-Harrasi et al., 2018).

Frankincense has diverse uses by various cultures, and the demand for resin extraction has increased considerably, which in turn has had negative impacts on the tree and its population, as noted in Ethiopia (Mengistu et al., 2013). This unsustainable harvest is greatly influencing the tree physiology and regeneration capacity (Mengistu et al., 2013), in which the number of cuts needed per tree to obtain resin is higher and higher for every season (Eshete, Sterck, & Bongers, 2012; Kebede, 2010; A. L. Khan et al., 2016). Anthropogenic tapping can enhance primary and secondary defence signalling in a variety of woody trees, for example, pine trees (Hofstetter, Mahfouz, Klepzig, & Ayres, 2005; Y.-B. Kim et al., 2009; Kramer, 2012; Langenheim, 2003; Trapp & Croteau, 2001). However, this subject has barely been studied in *Boswellia* species. In spite of the studies performed on *B. sacra* and *B. papyrifera* (Eshete et al., 2012; Mengistu et al., 2013; Tolera, Sass-Klaassen, Eshete, Bongers, & Sterck, 2013), tapping-induced physiochemical responses have not been fully studied in *Boswellia* species.

Boswellia Physiochemical Responses to Tapping

Effect of Wounding on Essential Nutrients and Amino Acid Accumulation

Ca^{2+} oscillation is known as a major signal that occurs during physical injury or related stressful conditions (Guo, Nazim, Liang, & Yang, 2016). The Ca^{2+} content is often regulated by plant responses to abiotic or biotic stress (Choi et al., 2017; Koldenkova & Hatsugai, 2018). Notably, calcium plays a key role in stomatal closure, cell proliferation, cell wall production, defence-related signalling and resistance against the negative effects of stress (Lautner and Fromm, 2010). A recent study by (A. L. Khan et al., 2018) showed that the *B. sacra* tree responded with significantly higher Ca^{2+} contents after 24 h of wounding stress to the tree compared to non-wounded control trees. This signal was also coupled with high amounts of magnesium as well, which is completely consistent with reports by Zebelo and Maffei (2014). A similar conclusion was also drawn regarding increased macronutrient accumulation, e.g. of phosphorus, potassium and nitrogen, which could be a result of the increased transpiration rate during wounding stress as shown by Schulze and Ehleringer (1984) during xylem tapping (Schulze & Ehleringer, 1984).

In addition to nutrients, physical injuries also impact the major building blocks such as amino acids and essential sugars. Most plants accumulate sugars in the form of nonstructural carbohydrates, which in turn reflects the carbon source and sink relationship (Bansal & Germino, 2009; Mengistu et al., 2013). Similar to other trees such as rubber from tropical and deciduous species, *Boswellia* species also showed an increased accumulation of carbon during wounding (A. L. Khan et al., 2018), suggesting a carbon sink strength or carbon supply relationship (Silpi et al., 2006). This finding was reported recently (A. L. Khan et al., 2018), in which a high concentration of carbon was accumulated. Synergistically, with increased wounding stress, essential amino acids such proline, sarcosine and GABA (associated with the cell wall; Fig. 4.2) will increase, and some will also act as osmoprotectants (Bown & Shelp, 2016; Scholz, Reichelt, Mekonnen, Ludewig, & Mithöfer, 2015).

Boswellia Tapping Influences Its Carbohydrate Metabolism

Plants store their major energy reservoirs in the form of carbohydrates and sugars, which are part of the Krebs cycle in photosynthesis (Silpi et al., 2007). A study was performed by Mengistu et al. (2013) on *B. papyrifera* that was growing wild as two major populations in Metemma and Abergele, Ethiopia. The authors selected two seasons, (i) wet and (ii) dry, and two tapping sessions, (i) 7 days and (ii) 14 days. After the treatments, the total soluble carbohydrates, total nonstructural carbohydrates and starch contents were measured using high-performance liquid chromatography (HPLC). The results showed that the season, tapping and specific tissues

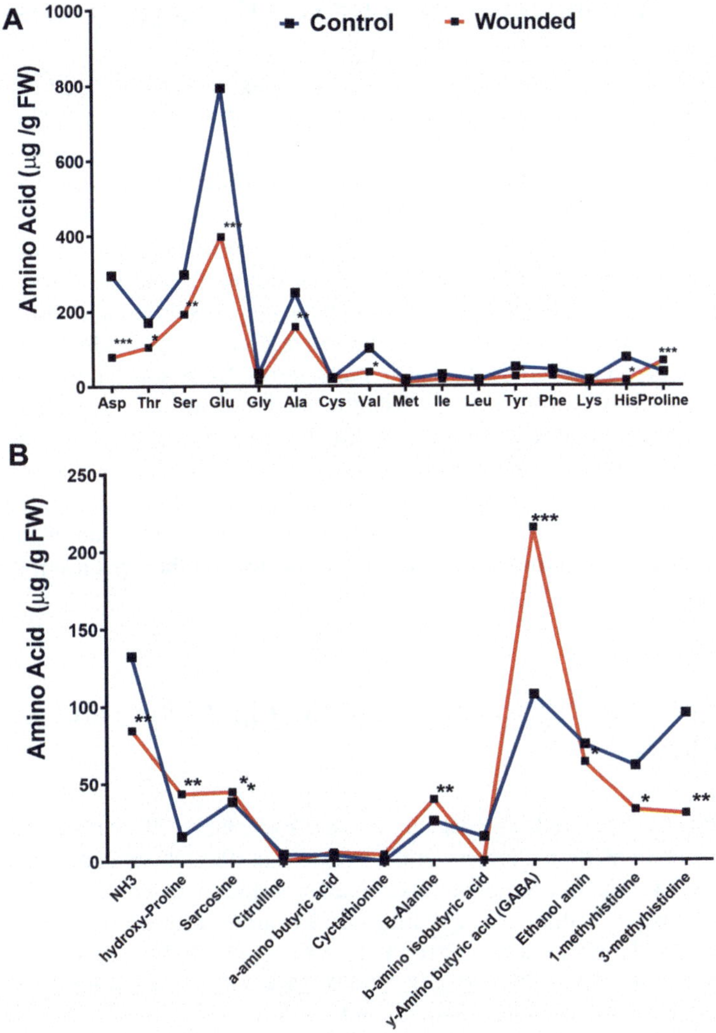

Fig. 4.2 The effects of wounding on changes in endogenous amino acids and osmolytes from *Boswellia sacra* samples

are significantly interdependent in correlation with carbohydrates and starch, whereas the siting of the trees was a non-significant determinant in this case (Mengistu et al., 2013). Similarly, these constituents were also significant in the highlands in the bark of the trees, while in the lowland population, these constituents were significant in the stem wood part. This study concludes that tapping significantly impacts tree carbon metabolism and causes a depletion in carbohydrates (Mengistu et al., 2013). In addition, the influence on carbon storage is directly proportional to the tree's growth and regeneration capacity.

Boswellia Tapping Influences Leaf Gas Exchange Processes

Boswellia growing in extreme conditions is often exposed to either a lower amount of water or a higher amount during the wet season. This tree has to adjust its internal physiological apparatus to cope with these imbalances from time to time. These conditions also diminish the chances that a tree will maintain a steady growth rate (Campanello, Gatti, & Goldstein, 2008). However, during the dry season, the tree sheds its leaves but maintains them under wet conditions, maintaining its water-carbon ratio (N. McDowell et al., 2008). In addition, the leaf is subject to signals to change the stomata closure and conductance. A study on diurnal leaf gas exchange was performed in Metemma and Abergele, Ethiopia, which represent the highland and lowland ecosystems of *B. papyrifera*. There was a population size of 90 trees in each location, with a substantial increase in leaf gas exchange in the morning and a gradual reduction during the day. Simultaneously, the transpiration also increased with increasing vapour pressure, thereby changing the stomatal closure. The gas exchange is limited in the low altitude population compared with the higher altitude one. Further studies will certainly assist in understanding the woody tree physiology. This approach can also be replicated in other species of this genus, such *B. sacra*.

Tapping Causes the Regulation of the Lipid Layer and JA Biosynthesis

As previously explained, physical injury leads to plant cell death and renders it prone to infections by pathogenic fungi or bacteria. Phytohormonal signalling is a key response for the tree to defend itself against further damage. Jasmonic acid (JA) has been the key to initiating a signalling cascade along with the lipid bilayer and its precursors (A. L. Khan et al., 2016; Y.-B. Kim et al., 2009; Y. Kim et al., 2018; Wasternack et al., 2006; Zarate et al., 2007). JA synthesis is a lipid-derived signal that is initiated by octadecanoids in the chloroplast, in which *12*-oxo-phytodienoic acid is the primary precursor of JA (Zarate et al., 2007). The first step of the biosynthetic reaction is when 13-lipoxygenase (13-LOX) catalyses the insertion of molecular oxygen into position 13 of α-linolenic acid, which is most likely released from the plastid membrane. Then, (13-S)-hydroperoxy linolenic acid (13-HPOT) is changed by allene oxide synthase (AOS), which is specific for changing 13-HPOT into an unstable allene oxide. This product is further processed as allene oxide cyclase (AOC). In the AOC-catalysed reaction, *cis*-(+)-OPDA is formed and then converts the enantiomeric structure at the cyclopentenone ring of the final product, (+)-7-iso-JA (Wasternack et al., 2006); (Figs. 4.3 and 4.4).

Similarly, fatty acids, such as polyunsaturated ones, are produced as a local response at the sites of infections or wounds. The first signal is received by lipases

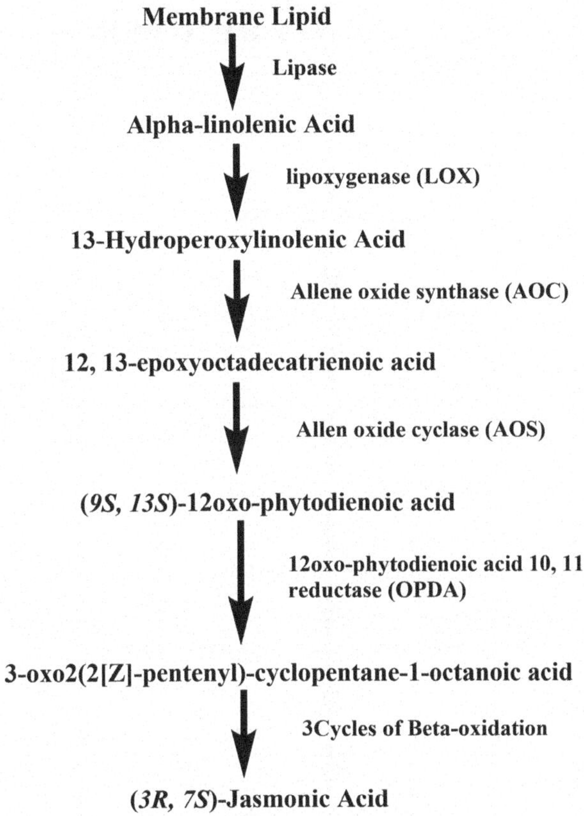

Fig. 4.3 Jasmonic acid biosynthesis pathway in plants

associated with cell membranes. It is a known substrate for lipoxygenase, which leads to the production of a lipid hydroperoxyl radical and leads to further signalling to synthesize various precursors and the primary signalling compounds such as wound-induced jasmonic acids. During wounding, LOOH is reduced to a lipid alkoxyl radical and forms polyunsaturated fatty acids (such as lipid radical $L^{\bullet}$; lipid hydroxide). All these reactions lead to the synthesis of JA as a final product. A recent study showed that the γ-linolenic acid content during the wounding of frankincense tree samples was significantly higher than that of non-wounded trees (Y. Kim et al., 2018). However, the contents of palmitic, vaccenic and linoleic acid were also significantly higher in the control trees (Fig. 4.5), whereas the stearic acid contents of the two treatments were nearly the same in both populations (Fig. 4.5). The JA was significantly higher in the wounded tree samples. This finding is also consistent with previous reports suggesting the activation of JA occurred in response to wounding stress (Fig. 4.5).

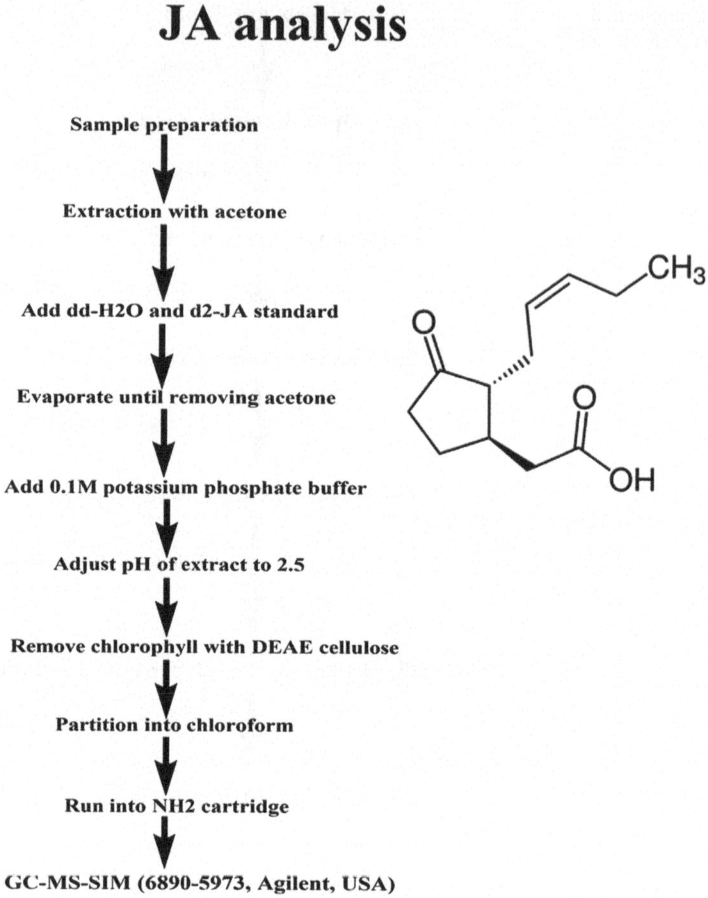

Fig. 4.4 JA analysis performed for *Boswellia* species

Effects of Incisions on Endogenous Salicylic Acid Regulation

Salicylic acid (SA) is also a defence-orchestrated hormone, similar to JA. However, most of its responses and its signalling cascade are triggered during pathogenic attack by developing constitutive systemic acquired resistance (SAR) (D. S. Kim & Hwang, 2014; Y. Kim et al., 2018). SA can regulate various physiological and biochemical responses such as the seed germination (Shakirova et al., 2003), photosynthesis pigment content (Fariduddin et al., 2003), transpiration rate, stomatal conductance and internal CO_2 concentration (Sh Hayat et al., 2010; M. A. Khan, Ungar, & Showalter, 2000), nitrate reductase activity (S Hayat et al., 2005), regulation of endogenous hormones (Sh Hayat et al., 2010; Shakirova et al., 2003) and modulation of antioxidant activity (Sh Hayat et al., 2010; Zou et al., 2009). Only a

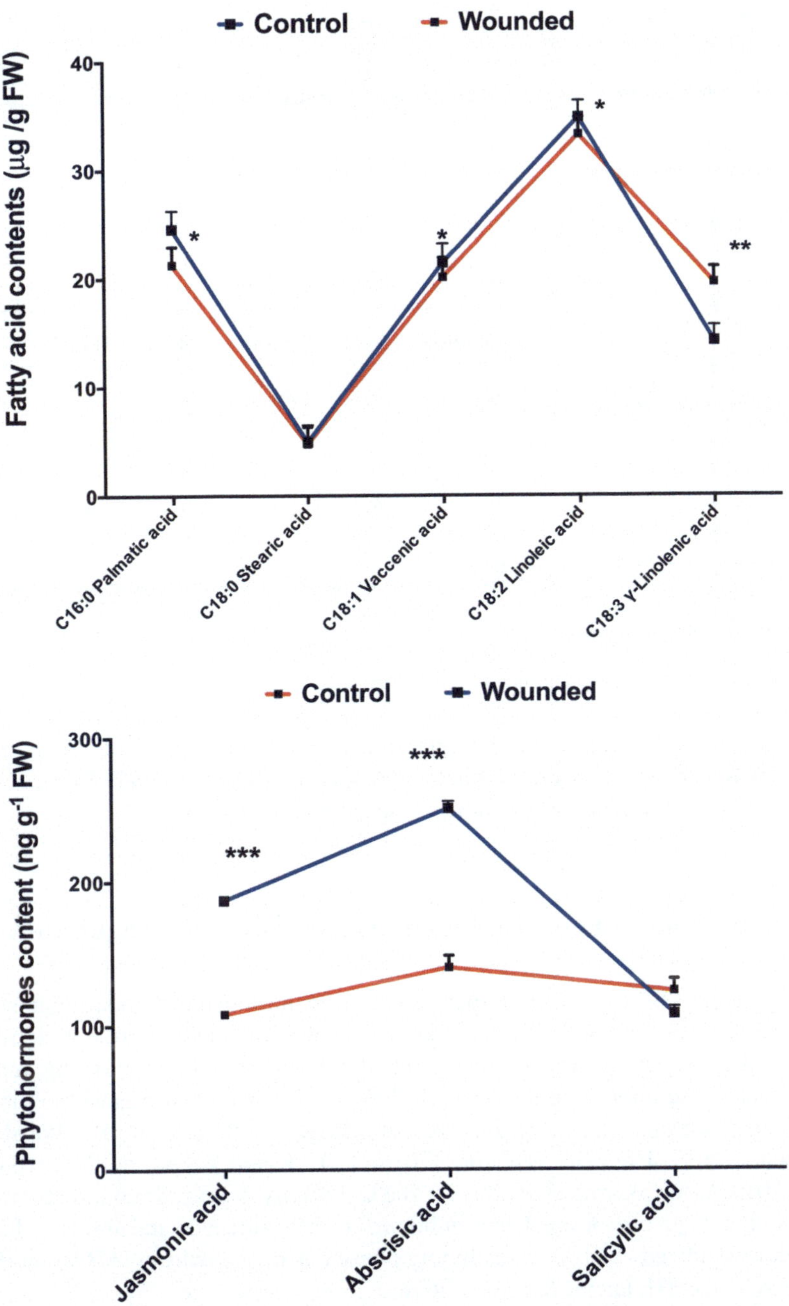

Fig. 4.5 Fatty acid degradation and endogenous phytohormone (jasmonic acid, abscisic acid and salicylic acid) analysis of *Boswellia sacra* trees in response to wounding (data adopted from (A. L. Khan et al., 2018)

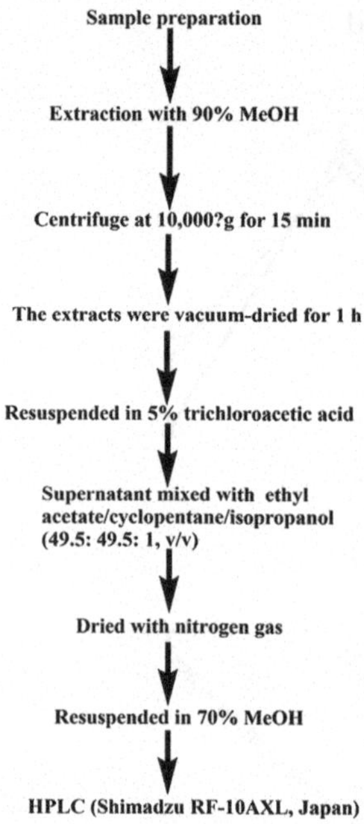

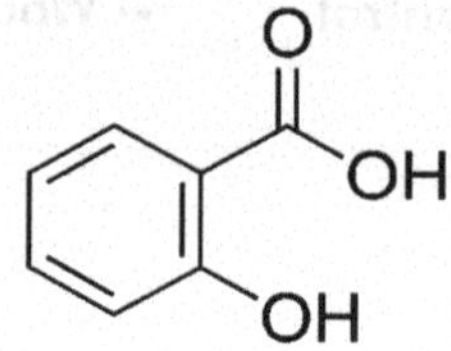

HPLC conditions

Gradient	5min	2.5min	4.5min	5min	3min
Solvent A	30%	40%	60%	30%	30%
Solvent B	70%	60%	40%	70%	70%

Fig. 4.6 SA extraction and quantification from *Boswellia* species

single study on *Boswellia* and SA quantification was performed by (A. L. Khan et al., 2018). The results showed that the SA was significantly greater in the control trees than in the wounded tree samples (Figs. 4.5 and 4.6).

Effect of Wounding on Endogenous Abscisic Acid Regulation

Abscisic acid (ABA) is a sesquiterpene and a stress hormone in plants, and it is usually abundant in terminal buds and is known to play an epicentric role in dehydration, seed germination, fruit development, floral initiation and combating various stress factors (salinity, drought, heat, chilling, heavy metal and organic pollutants, pathogenic attacks, etc.) (Dekkers & Smeekens, 2018; Hauser, Li, Waadt, & Schroeder, 2017; Helander & Cutler, 2018; A. L. Khan, Waqas, et al., 2015a; Qi et al., 2014; Vishwakarma et al., 2017). The C-15 ABA skeleton is commonly found in biosynthetic precursors such as xanthoxin, abscisic aldehyde and abscisic alcohol as well as oxidized catabolites including phaseic acid, 8′-hydroxy-ABA and dihydrophaseic acid (Helander & Cutler, 2018).

In the case of *Boswellia*, little is known about the stress aversion and regulation of ABA. (Al-Harrasi et al., 2018) reported that after a wounding stress lasting 24 h in the *B. sacra* tree, the stress response causes a significantly higher amount of ABA synthesis compared to trees that were not wounded. These analyses were performed

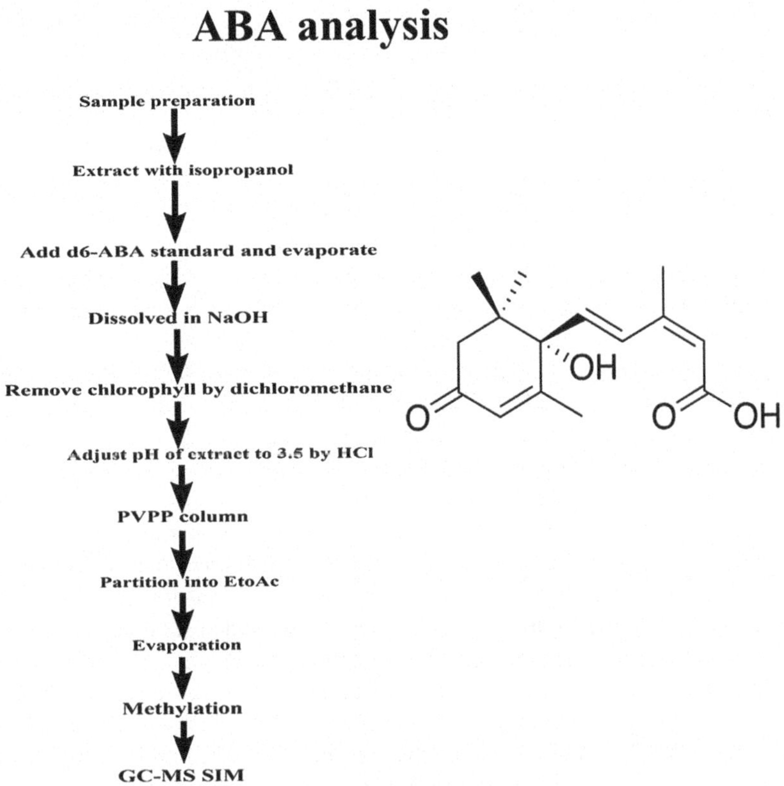

Fig. 4.7 ABA analysis of the leaf portion as performed for *Boswellia sacra*

through extraction, column chromatography and gas chromatography-mass spectrophotometry coupled with selected ion monitor analysis (Fig. 4.7). However, these analyses were performed on the leaf responses, and ways in which the stem responds to this localized stress are not yet known and require further study.

Enhanced Elicitation of Endogenous GA in *Boswellia*

Gibberellins (GAs), a class of novel hormone and diterpenoids, are small signalling molecules that are efficient and effective at minute concentrations. They can influence cell development and division processes, seed growth, sex, flowering, fruit development and stress responses (Bömke & Tudzynski, 2009; Hedden & Sponsel, 2015; Hedden & Thomas, 2012; A. L. Khan et al., 2016; A. L. Khan et al., 2012; A. L. Khan, Waqas, et al., 2015a; MacMillan, 2001; Thomas, 2018). There are 135 different gibberellins (GAs) that were first identified from *Gibberella fujikuroi* (Mihlan, Homann, Liu, & Tudzynski, 2003). That study was further extended by

Takahashi et al. (1955) as GA1, GA2 and GA3 caused a stir in the scientific community, leading to work on plant growth hormones to initiate the green revolution. More information on GA can be found at http://www.plant-hormones.info/gibberellin_nomenclature.htm. Among the GAs, only GA1, GA3, GA4, GA7 and GA20 are physiologically active, whereas the functions of the other hormones are not known, but they are involved in the biosynthesis pathways as precursors or intermediates.

GAs have been found to be produced not only by plants but more recently by various microbes (fungi and bacteria) and some insects as well (A. L. Khan et al., 2016; A. L. Khan et al., 2012; A. L. Khan, Hussain, et al., 2015). In the *Boswellia* species, only the GA of *B. sacra* has been addressed through the present, and there is meagre information on this subject. The results in Shahzad et al. (2016) showed that the endogenous GAs were significantly higher in wounded trees than unwounded ones, which supports the role of their growth-regulating potentials. This finding suggests that during wounding stress, the tree activated its cell division during resin secretion to avoid compromising tissue death. However, further studies are essential for understanding these growth dynamics. Other GAs were not significantly different in their contents compared to the control tree during wounding stress.

The GA_{12} content of the control samples was higher than that of the wounded tree (A. L. Khan et al., 2018). A similar trend was observed for the GA_{19} content, which was 4.32% higher in the control than in the wounded samples. However, the later stages of GA biosynthesis were significantly more active in the wounded trees than in the control trees, showing the intuitive response of the trees towards wounding. The GA_3 content was nearly sevenfold greater in the wounded trees than in the control trees. The GA_1 contents of the two populations were significantly different. The GA_{20} content was significantly higher in the wounded tree samples. Similarly, the GA_4 content was higher in the wounded tree samples than in the control tree samples. In relation to the gibberellins (GAs), GA_3, GA_{20} and GA_4 were significantly more activated in the wounded trees than in the control trees, whereas GA_{12}, GA_{19} and GA_9 were significantly less activated.

Despite the analysis of three replicates, no GA_{53}, GA_{34} or GA_8 were detected in either of the populations. GA_9 was only found in the wounded trees, whereas physiologically inactive (GA_{12}, GA_{19} and GA_{20}) and physiologically active (GA_1, GA_4 and GA_3) GAs were detected in both the wounded and preserved tree samples. This result suggests the existence of both C-13 and non-C-13 hydroxylation pathways in *B. sacra*, and GA could be synthesized as follows: *ent*-kaurenoic acid → GA_{12} aldehyde → GA_{12} → GA_{19}, → GA_{20} → GA_1 → GA_3 (Fig. 4.8; Bilal et al., 2018; A. L. Khan et al., 2018; Rehman et al., 2018) (Bömke & Tudzynski, 2009; A. L. Khan, Waqas, et al., 2015b; Thomas, 2018).

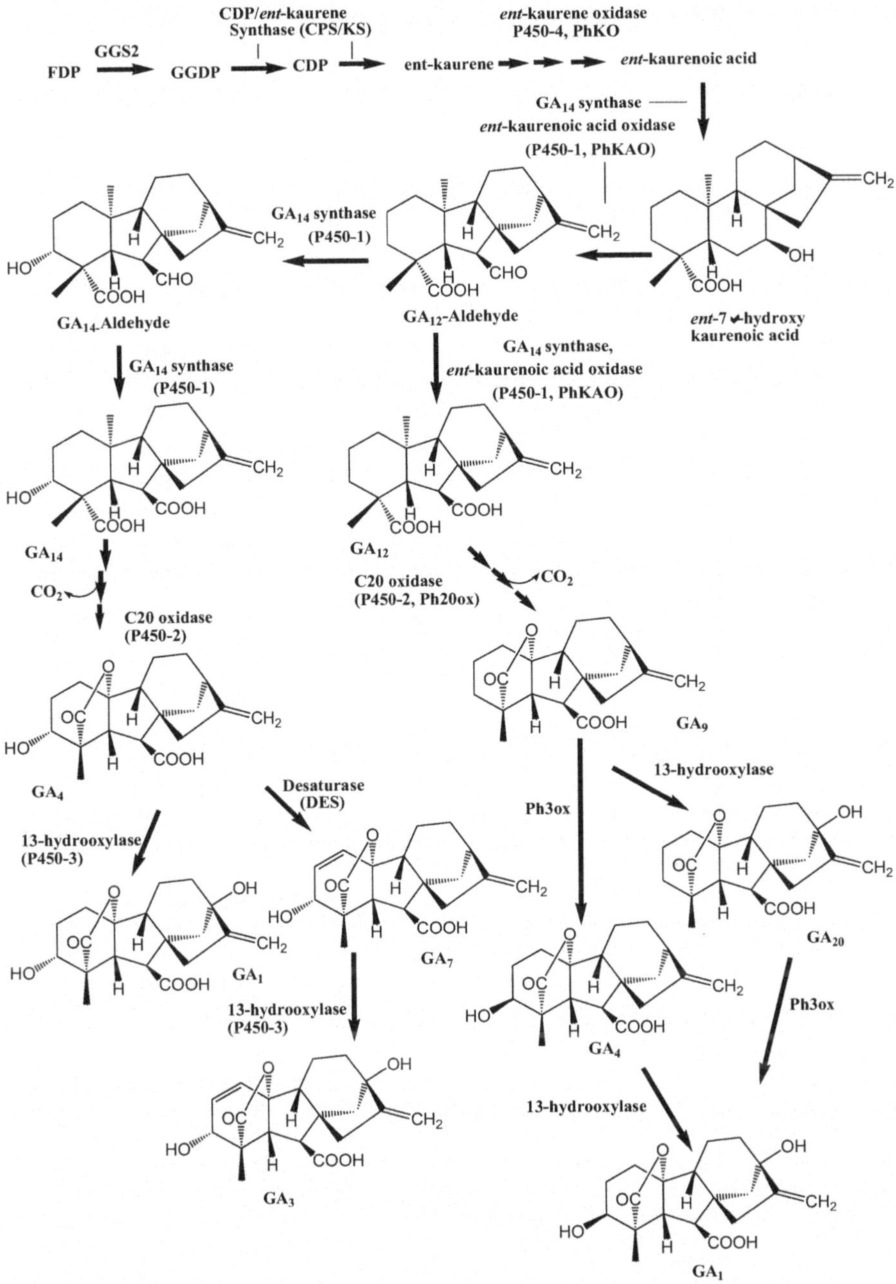

Fig. 4.8 The gibberellin biosynthesis pathways in plants

Boswellia Gene Expression Patterns During Wounding

The gene expression of the GA biosynthesis-related gene ent-copalyl diphosphate synthase (CPS), which catalyses geranylgeranyl diphosphate to ent-copalyl diphosphate (Prisic, Xu, Wilderman, & Peters, 2004), was upregulated during wounding stress relative to the control samples. At the same time, the cytochrome P450 monooxygenases CyP450-P450 were responsible for catalysing the 16α,17-epoxidation of non-13-hydroxylated gibberellins (GAs), including GA_4, GA_9 and GA_{12} (Magome et al., 2013; Zhu et al., 2006), and GA_{20} oxidase (GA_{12} and GA_{53} to GA_9 and GA_{20}) through a three-step oxidation at C-20 in the GA structure (Rieu et al., 2008). This finding shows the highest amount of gene expression as quantified through qRT-PCR analysis and was compared with two housekeeping genes specifically designed for *Boswellia sacra* (A. L. Khan et al., 2018).

In addition, both *allene oxide cyclase* (AOC) and *allene oxide synthase* (AOS) were more significantly expressed in tapped trees than in control trees. The AOS gene is directly related to the JA biosynthesis pathway, and thus, wound induction is elicited through the activation of the JA biosynthesis pathway (Lacombe & Achard, 2016; Suzuki et al., 2016). Further detailed studies of RNA transcriptomics may improve our understanding of *B. sacra* wounding.

References

Al-Harrasi, A., Rehman, N. U., Khan, A. L., Al-Broumi, M., Al-Amri, I., Hussain, J., … Csuk, R. (2018). Chemical, molecular and structural studies of Boswellia species: β-Boswellic Aldehyde and 3-epi-11β-Dihydroxy BA as precursors in biosynthesis of boswellic acids. *PLoS One, 13*(6), e0198666.

Bansal, S., & Germino, M. J. (2009). Temporal variation of nonstructural carbohydrates in montane conifers: Similarities and differences among developmental stages, species and environmental conditions. *Tree Physiology, 29*(4), 559–568.

Bilal, S., Khan, A. L., Shahzad, R., Kim, Y.-H., Imran, M., Khan, M. J., … Lee, I.-J. (2018). Mechanisms of Cr (VI) resistance by endophytic Sphingomonas sp. LK11 and its Cr (VI) phytotoxic mitigating effects in soybean (*Glycine max* L.). *Ecotoxicology and Environmental Safety, 164*, 648–658.

Bömke, C., & Tudzynski, B. (2009). Diversity, regulation, and evolution of the gibberellin biosynthetic pathway in fungi compared to plants and bacteria. *Phytochemistry, 70*(15–16), 1876–1893.

Bown, A. W., & Shelp, B. J. (2016). Plant GABA: not just a metabolite. *Trends in Plant Science, 21*(10), 811–813.

Campanello, P. I., Gatti, M. G., & Goldstein, G. (2008). Coordination between water-transport efficiency and photosynthetic capacity in canopy tree species at different growth irradiances. *Tree Physiology, 28*(1), 85–94.

Choi, W. G., Miller, G., Wallace, I., Harper, J., Mittler, R., & Gilroy, S. (2017). Orchestrating rapid long-distance signaling in plants with Ca2+, ROS and electrical signals. *The Plant Journal, 90*(4), 698–707.

Davey, M., Stals, E., Panis, B., Keulemans, J., & Swennen, R. (2005). High-throughput determination of malondialdehyde in plant tissues. *Analytical Biochemistry, 347*(2), 201–207.

Dekkers, B. J., & Smeekens, S. C. (2018). Sugar and abscisic acid regulation of germination and transition to seedling growth. *Annual Plant Reviews, 27*, 305–327.

Eshete, A., Sterck, F. J., & Bongers, F. (2012). Frankincense production is determined by tree size and tapping frequency and intensity. *Forest Ecology and Management, 274*, 136–142.

Farah, A. Y. (1994). *The milk of the Boswellia forests: frankincense production among the pastoral Somali*: EPOS, Environmental Policy and Society, Uppsala, Sweden.

Fariduddin, Q., Hayat, S., & Ahmad, A. (2003). Salicylic acid influences net photosynthetic rate, carboxylation efficiency, nitrate reductase activity, and seed yield in *Brassica juncea. Photosynthetica, 41*(2), 281–284.

Farmer, E. E., Alméras, E., & Krishnamurthy, V. (2003). Jasmonates and related oxylipins in plant responses to pathogenesis and herbivory. *Current Opinion in Plant Biology, 6*(4), 372–378.

Guo, W., Nazim, H., Liang, Z., & Yang, D. (2016). Magnesium deficiency in plants: An urgent problem. *The Crop Journal, 4*(2), 83–91.

Hauser, F., Li, Z., Waadt, R., & Schroeder, J. I. (2017). SnapShot: abscisic acid signaling. *Cell, 171*(7), 1708–1708. e1700.

Hayat, S., Fariduddin, Q., Ali, B., & Ahmad, A. (2005). Effect of salicylic acid on growth and enzyme activities of wheat seedlings. *Acta Agronomica Hungarica, 53*(4), 433–437.

Hayat, S., Yadav, S., Ali, B., & Ahmad, A. (2010). Interactive effect of nitric oxide and brassinosteroids on photosynthesis and the antioxidant system of *Lycopersicon esculentum. Russian Journal of Plant Physiology, 57*(2), 212–221.

Hedden, P., & Sponsel, V. (2015). A century of gibberellin research. *Journal of Plant Growth Regulation, 34*(4), 740–760.

Hedden, P., & Thomas, S. G. (2012). Gibberellin biosynthesis and its regulation. *Biochemical Journal, 444*(1), 11–25.

Helander, J. D., & Cutler, S. R. (2018). Abscisic acid signaling and biosynthesis: Protein structures and molecular probes. In *Plant Structural Biology: Hormonal Regulations* (pp. 113–146). Springer, Cham.

Hofstetter, R., Mahfouz, J. B., Klepzig, K. D., & Ayres, M. (2005). Effects of tree phytochemistry on the interactions among endophloedic fungi associated with the southern pine beetle. *Journal of Chemical Ecology, 31*(3), 539–560.

Holbrook, N. M., & Knoblauch, M. (2018). *Editorial overview: Physiology and metabolism: Phloem: a supracellular highway for the transport of sugars, signals, and pathogens*: Current Opinion in Plant Biology, 43, iii-vii.

Hyodo, H. (2017). *Stress/wound ethylene The plant hormone ethylene* (pp. 43-63): CRC Press, Florida, USA.

Jacobo-Velázquez, D. A., González-Agüero, M., & Cisneros-Zevallos, L. (2015). Cross-talk between signaling pathways: The link between plant secondary metabolite production and wounding stress response. *Scientific Reports, 5*, 8608.

Kebede, T. (2010). *Current production systems of frankincense from Boswellia papyrifera tree; its implications on sustainable utilization of the resource.* MSc thesis, Mekelle University, Mekelle.

Khan, A. L., Al-Harrasi, A., Al-Rawahi, A., Al-Farsi, Z., Al-Mamari, A., Waqas, M., … Shin, J.-H. (2016). Endophytic fungi from frankincense tree improves host growth and produces extracellular enzymes and indole acetic acid. *PLoS One, 11*(6), e0158207.

Khan, A. L., Al-Harrasi, A., Shahzad, R., Imran, Q. M., Yun, B.-W., Kim, Y.-H., … Lee, I.-J. (2018). Regulation of endogenous phytohormones and essential metabolites in frankincense-producing *Boswellia sacra* under wounding stress. *Acta Physiologiae Plantarum, 40*(6), 113.

Khan, A. L., Hamayun, M., Kang, S.-M., Kim, Y.-H., Jung, H.-Y., Lee, J.-H., & Lee, I.-J. (2012). Endophytic fungal association via gibberellins and indole acetic acid can improve plant growth under abiotic stress: An example of Paecilomyces formosus LHL10. *BMC Microbiology, 12*(1), 3.

Khan, A. L., Hussain, J., Al-Harrasi, A., Al-Rawahi, A., & Lee, I.-J. (2015). Endophytic fungi: Resource for gibberellins and crop abiotic stress resistance. *Critical Reviews in Biotechnology, 35*(1), 62–74.

Khan, A. L., Waqas, M., Asaf, S., Kamran, M., Shahzad, R., Bilal, S., ... Yun, B.-W. (2017). Plant growth-promoting endophyte Sphingomonas sp. LK11 alleviates salinity stress in *Solanum pimpinellifolium*. *Environmental and Experimental Botany, 133*, 58–69.

Khan, A. L., Waqas, M., Hussain, J., Al-Harrasi, A., Hamayun, M., & Lee, I.-J. (2015a). Phytohormones enabled endophytic fungal symbiosis improve aluminum phytoextraction in tolerant *Solanum lycopersicum*: An examples of Penicillium janthinellum LK5 and comparison with exogenous GA3. *Journal of Hazardous Materials, 295*, 70–78.

Khan, A. L., Waqas, M., Hussain, J., Al-Harrasi, A., Hamayun, M., & Lee, I.-J. (2015b). Phytohormones enabled endophytic fungal symbiosis improve aluminum phytoextraction in tolerant *Solanum lycopersicum*: An examples of Penicillium janthinellum LK5 and comparison with exogenous GA 3. *Journal of Hazardous Materials, 295*, 70–78.

Khan, M. A., Ungar, I. A., & Showalter, A. M. (2000). Effects of sodium chloride treatments on growth and ion accumulation of the halophyte Haloxylon recurvum. *Communications in Soil Science & Plant Analysis, 31*(17–18), 2763–2774.

Kim, D. S., & Hwang, B. K. (2014). An important role of the pepper phenylalanine ammonia-lyase gene (PAL1) in salicylic acid-dependent signalling of the defence response to microbial pathogens. *Journal of Experimental Botany, 65*(9), 2295–2306.

Kim, I. S., Koppula, S., Park, S. Y., & Choi, D. K. (2017). Analysis of epidermal growth factor receptor related gene expression changes in a cellular and animal model of Parkinson's disease. *International Journal of Molecular Sciences, 18*(2). https://doi.org/10.3390/ijms18020430

Kim, Y.-B., Kim, S.-M., Kang, M.-K., Kuzuyama, T., Lee, J. K., Park, S.-C., ... Kim, S.-U. (2009). Regulation of resin acid synthesis in *Pinus densiflora* by differential transcription of genes encoding multiple 1-deoxy-D-xylulose 5-phosphate synthase and 1-hydroxy-2-methyl-2-(E)-butenyl 4-diphosphate reductase genes. *Tree Physiology, 29*(5), 737–749.

Kim, Y.-H., Hwang, S.-J., Waqas, M., Khan, A. L., Lee, J.-H., Lee, J.-D., ... Lee, I.-J. (2015). Comparative analysis of endogenous hormones level in two soybean (*Glycine max* L.) lines differing in waterlogging tolerance. *Frontiers in Plant Science, 6*, 714.

Kim, Y., Seo, C.-W., Khan, A. L., Mun, B.-G., Shahzad, R., Ko, J.-W., ... Lee, I.-J. (2018). Ethylene mitigates waterlogging stress by regulating glutathione biosynthesis-related transcripts in soybeans. *bioRxiv*, 252312.

Klay, I., Gouia, S., Liu, M., Mila, I., Khoudi, H., Bernadac, A., ... Pirrello, J. (2018). Ethylene Response Factors (ERF) are differentially regulated by different abiotic stress types in tomato plants. *Plant Science, 274*, 137.

Koldenkova, V. P., & Hatsugai, N. (2018). How do Plants Keep their Functional Integrity? *Plant signaling & behavior, 13*(8), e1464853.

Koo, A. J. (2018). Metabolism of the plant hormone jasmonate: A sentinel for tissue damage and master regulator of stress response. *Phytochemistry Reviews, 17*(1), 51–80.

Kovalchuk, A., Keriö, S., Oghenekaro, A. O., Jaber, E., Raffaello, T., & Asiegbu, F. O. (2013). Antimicrobial defenses and resistance in forest trees: Challenges and perspectives in a genomic era. *Annual Review of Phytopathology, 51*, 221–244.

Kramer, P. (2012). *Physiology of woody plants*: Academic Press, New York, USA.

Lacombe, B., & Achard, P. (2016). Long-distance transport of phytohormones through the plant vascular system. *Current Opinion in Plant Biology, 34*, 1–8.

Langenheim, J. H. (2003). *Plant resins: Chemistry, evolution, ecology, and ethnobotany*. Oregon, US: Timber Press.

Lautner, S., & Fromm, J. (2010). Calcium-dependent physiological processes in trees. *Journal of Plant Biology, 12*(2), 268–274.

León, J., Rojo, E., & Sánchez-Serrano, J. J. (2001). Wound signalling in plants. *Journal of Experimental Botany, 52*(354), 1–9.

MacMillan, J. (2001). Occurrence of gibberellins in vascular plants, fungi, and bacteria. *Journal of Plant Growth Regulation, 20*(4), 387–442.

Magome, H., Nomura, T., Hanada, A., Takeda-Kamiya, N., Ohnishi, T., Shinma, Y., ... Yamaguchi, S. (2013). CYP714B1 and CYP714B2 encode gibberellin 13-oxidases that reduce gibberellin activity in rice. *Proceedings of the National Academy of Sciences, 110*(5), 1947–1952.

McDowell, N., Pockman, W. T., Allen, C. D., Breshears, D. D., Cobb, N., Kolb, T., … Williams, D. G. (2008). Mechanisms of plant survival and mortality during drought: Why do some plants survive while others succumb to drought? *New Phytologist, 178*(4), 719–739.

McDowell, R. E., Amsler, M. O., Li, Q., Lancaster, J. R., Jr., & Amsler, C. D. (2015). The immediate wound-induced oxidative burst of Saccharina latissima depends on light via photosynthetic electron transport. *Journal of Phycology, 51*(3), 431–441.

Mengistu, T., Sterck, F. J., Anten, N. P., & Bongers, F. (2012). Frankincense tapping reduced photosynthetic carbon gain in *Boswellia papyrifera* (Burseraceae) trees. *Forest Ecology and Management, 278*, 1–8.

Mengistu, T., Sterck, F. J., Fetene, M., & Bongers, F. (2013). Frankincense tapping reduces the carbohydrate storage of Boswellia trees. *Tree Physiology, 33*(6), 601–608.

Mihlan, M., Homann, V., Liu, T. W. D., & Tudzynski, B. (2003). AREA directly mediates nitrogen regulation of gibberellin biosynthesis in Gibberella fujikuroi, but its activity is not affected by NMR. *Molecular Microbiology, 47*(4), 975–991.

Oyarce, P., & Gurovich, L. (2011). Evidence for the transmission of information through electric potentials in injured avocado trees. *Journal of Plant Physiology, 168*(2), 103–108.

Prisic, S., Xu, M., Wilderman, P. R., & Peters, R. J. (2004). Rice contains two disparate ent-copalyl diphosphate synthases with distinct metabolic functions. *Plant Physiology, 136*(4), 4228–4236.

Qi, X., Li, M.-W., Xie, M., Liu, X., Ni, M., Shao, G., … Wong, F.-L. (2014). Identification of a novel salt tolerance gene in wild soybean by whole-genome sequencing. *Nature Communications, 5*, 4340.

Rehman, N. U., Ali, L., Al-Harrasi, A., Mabood, F., Al-Broumi, M., Khan, A. L., … Csuk, R. (2018). Quantification of AKBA in *Boswellia sacra* using NIRS coupled with PLSR as an alternative method and cross-validation by HPLC. *Phytochemical Analysis, 29*(2), 137–143.

Rieu, I., Ruiz-Rivero, O., Fernandez-Garcia, N., Griffiths, J., Powers, S. J., Gong, F., … Thomas, S. G. (2008). The gibberellin biosynthetic genes AtGA20ox1 and AtGA20ox2 act, partially redundantly, to promote growth and development throughout the Arabidopsis life cycle. *The Plant Journal, 53*(3), 488–504.

Rijkers, T., Ogbazghi, W., Wessel, M., & Bongers, F. (2006). The effect of tapping for frankincense on sexual reproduction in *Boswellia papyrifera*. *Journal of Applied Ecology, 43*(6), 1188–1195.

Savatin, D. V., Gramegna, G., Modesti, V., & Cervone, F. (2014). Wounding in the plant tissue: The defense of a dangerous passage. *Frontiers in Plant Science, 5*, 470.

Scholz, S. S., Reichelt, M., Mekonnen, D. W., Ludewig, F., & Mithöfer, A. (2015). Insect herbivory-elicited GABA accumulation in plants is a wound-induced, direct, systemic, and jasmonate-independent defense response. *Frontiers in Plant Science, 6*, 1128.

Schulze, E.-D., & Ehleringer, J. (1984). The effect of nitrogen supply on growth and water-use efficiency of xylem-tapping mistletoes. *Planta, 162*(3), 268–275.

Shahzad, R., Waqas, M., Khan, A. L., Asaf, S., Khan, M. A., Kang, S.-M., … Lee, I.-J. (2016). Seed-borne endophytic *Bacillus amyloliquefaciens* RWL-1 produces gibberellins and regulates endogenous phytohormones of *Oryza sativa*. *Plant Physiology and Biochemistry, 106*, 236–243.

Shakirova, F. M., Sakhabutdinova, A. R., Bezrukova, M. V., Fatkhutdinova, R. A., & Fatkhutdinova, D. R. (2003). Changes in the hormonal status of wheat seedlings induced by salicylic acid and salinity. *Plant Science, 164*(3), 317–322.

Silpi, U., Lacointe, A., Kasempsap, P., Thanysawanyangkura, S., Chantuma, P., Gohet, E., … Thaler, P. (2007). Carbohydrate reserves as a competing sink: Evidence from tapping rubber trees. *Tree Physiology, 27*(6), 881–889.

Silpi, U., Thaler, P., Kasemsap, P., Lacointe, A., Chantuma, A., Adam, B., … Améglio, T. (2006). Effect of tapping activity on the dynamics of radial growth of *Hevea brasiliensis* trees. *Tree Physiology, 26*(12), 1579–1587.

Suzuki, N., Bassil, E., Hamilton, J. S., Inupakutika, M. A., Zandalinas, S. I., Tripathy, D., … Kumazaki, A. (2016). ABA is required for plant acclimation to a combination of salt and heat stress. *PLoS One, 11*(1), e0147625.

Tadesse, W., Feleke, S., & Eshete, T. (2004). Comparative study of traditional and new tapping methods on frankincense yield of *boswellia papyrifera*. *Ethiopian Journal of Natural Resources*.

Takahashi, N., Kitamura, H., Kawarada, A., Seta, Y., Takai, M., Tamura, S., & Sumiki, Y. (1955). Biochemical studies on "Bakanae" fungus: Part XXXIV. Isolation of gibberellins and their properties part XXXV. Relation between gibberellins, A1, A2 and gibberellic acid. *Journal of the Agricultural Chemical Society of Japan, 19*(4), 267–281.

Thomas, S. G., & Hedden, P. (2018). Gibberellin metabolism and signal transduction. *Annual Plant Reviews, 24*, 147–184.

Tolera, M., Sass-Klaassen, U., Eshete, A., Bongers, F., & Sterck, F. J. (2013). Frankincense tree recruitment failed over the past half century. *Forest Ecology and Management, 304*, 65–72.

Torres-Contreras, A. M., Senés-Guerrero, C., Pacheco, A., González-Agüero, M., Ramos-Parra, P. A., Cisneros-Zevallos, L., & Jacobo-Velázquez, D. A. (2018). Genes differentially expressed in broccoli as an early and late response to wounding stress. *Postharvest Biology and Technology, 145*, 172–182.

Trapp, S. C., & Croteau, R. B. (2001). Genomic organization of plant terpene synthases and molecular evolutionary implications. *Genetics, 158*(2), 811–832.

Vishwakarma, K., Upadhyay, N., Kumar, N., Yadav, G., Singh, J., Mishra, R. K., . . . Pandey, M. (2017). Abscisic acid signaling and abiotic stress tolerance in plants: a review on current knowledge and future prospects. *Frontiers in plant science, 8*, 161–174.

Wasternack, C., Stenzel, I., Hause, B., Hause, G., Kutter, C., Maucher, H., ... Miersch, O. (2006). The wound response in tomato–role of jasmonic acid. *Journal of Plant Physiology, 163*(3), 297–306.

Zarate, S. I., Kempema, L. A., & Walling, L. L. (2007). Silverleaf whitefly induces salicylic acid defenses and suppresses effectual jasmonic acid defenses. *Plant Physiology, 143*(2), 866–875.

Zebelo, S. A., & Maffei, M. E. (2014). Role of early signalling events in plant–insect interactions. *Journal of Experimental Botany, 66*(2), 435–448.

Zhou, Y., & Underhill, S. J. (2015). Breadfruit (*Artocarpus altilis*) gibberellin 20-oxidase genes: Sequence variants, stem elongation and abiotic stress response. *Tree Genetics & Genomes, 11*(4), 84.

Zhu, Y., Nomura, T., Xu, Y., Zhang, Y., Peng, Y., Mao, B., ... Li, P. (2006). ELONGATED UPPERMOST INTERNODE encodes a cytochrome P450 monooxygenase that epoxidizes gibberellins in a novel deactivation reaction in rice. *The Plant Cell, 18*(2), 442–456.

Zou, J., Liu, A., Chen, X., Zhou, X., Gao, G., Wang, W., & Zhang, X. (2009). Expression analysis of nine rice heat shock protein genes under abiotic stresses and ABA treatment. *Journal of Plant Physiology, 166*(8), 851–861.

Chapter 5
Propagation and Conservation of *Boswellia sacra*

Ecology and Environmental Conditions

The socio-economic significance and ecological importance of *Boswellia* species are broadly acknowledged around the world. Various *Boswellia* species have been reported to grow under different environmental conditions within tropical zones. For example, *B. serrata* grows in the humid tropics, while *B. sacra* grows in dry tropical regions, and both species have adapted to the Sonoran Desert with its very short winter season. However, *B. socotrana* did not adapt to this region because even in a greenhouse environment, this species prefers the winter. It is not easy to make a simple decision and assumption about a suitable habitat for its growth. Every species has a specific favourable environment, and therefore it is necessary to experiment before drawing a conclusion and setting parameters for cultivating *Boswellia*. Eslamieh (2017) reported that almost all *Boswellia* species need humidity and heat, while there is no broad commonality among these species otherwise; each species has its own genetic makeup for cultivation and adaptation.

Environmental Effects on Plant Morphology

Plants are transformed into different forms due to the effects of different environmental conditions. These conditions significantly and constantly affect the plant structure, creating a unique shape. These natural environmental conditions can be difficult to arrange for the cultivation of some *Boswellia* species. Usually, reductions in leaf size and number, extra swollen bases and prostrate growth and many other morphological modifications are due to harsh environmental conditions. Various *Boswellia* species grown in different environments have different morphological appearances unlike those of their original habitat (Eslamieh, 2017). For example, when grown in rock crevasses, *B. nana* totally changed its shape to

A. Al-Harrasi et al., *Biology of Genus Boswellia*,
https://doi.org/10.1007/978-3-030-16725-7_5

fit the rock cavities, while in the field, it grows similar to a dwarf shrub. Another example was reported by Eslamieh (2017) in which *B. frereana* formed a disk-shaped mass in an extreme habitat, while its cultivated form grows similar to a normal plant.

Soil Preparation and Planting

Various *Boswellia* species grow under different environmental conditions. For example, some species can grow in the humid tropics and some grow in dry tropical regions. Due to the natural environmental conditions, the growth condition will change according to their respective species. For example, in the dry tropics, plants grew well in soil medium that had equal parts of quarter-inch solid granular material (decomposed granite clay balls, pumice and coarse sand or equivalent), screened perlite and organic material such as sterilized compost (Eslamieh, 2017). This mixture is more suitable for *Boswellia* species, prevents waterlogged conditions and helps the plant to obtain proper oxygen and nutrients. However, various researchers or growers have replaced perlite with vermiculite. Moreover, in the areas where pumice is not available, other materials can be used instead, such as granular gypsum, decomposed granite, clay balls and other rock-type bonsai mixes. In fact, the above-mentioned product can provide the essential nutrients required by most *Boswellia* species. However, it will be more useful for the plant if the grower uses a low pH (5–5.5) water such as rain water. Without rain water or low pH watering, the alkalinity of the mix will increase and eventually become harmful to the plant. Crusting is a phenomenon in which the topsoil of a container becomes hard from small particles that make a tight bond across the surface, which builds up a layer of salt and minerals around the root ball, prevents water flow and builds up excessive salt, leading to plant death.

Another concern is that the soil for this group requires a medium containing more organic material with proper drainage. The *Boswellia* species (humid-tropical) grow well during active months in the presence of proper moisture contents. However, under harsh conditions, such as those found in deserts, regular irrigation is required under straight compost, and growers should never allow the soil to dry during the active month for humid-tropical species (Eslamieh, 2017). As we discussed earlier, sufficient air and oxygen are very important for the root system. During the resting period, the soil should be allowed to dry out before watering, but the soil should never stay dry for longer than a week. *Boswellia* plants can survive for longer times when the soil is dry and there is minimum growth. The above-mentioned media are ideal for *Boswellia* growth. *Boswellia* species can grow in harsh environments for long periods of time without any serious adverse effects.

Irrigation of *Boswellia* Plants

A question now arises regarding how much water is required for *Boswellia* plant growth. The amount of water for growth varies from environment to environment and season to season. The amount and frequency of watering depends on various conditions and factors. A grower must consider certain factors when setting up an irrigation system for *Boswellia* plants, especially those in pots. *Boswellia* species lack an appropriate system for retaining water for longer times during harsh conditions because they are semi-succulents. However, they can develop tubers during possible drought seasons. A *Boswellia* plant can survive in a pot as well as in the field without water for some time, but after a few weeks, it will start defoliating. Supplying sufficient water after defoliation can lead to a quick recovery for the plants. *Boswellia* plants are favoured by having water running through their root systems if the soil drains well during the growing season, and they will develop a strong root system and vigorous growth.

Nutrient Requirements for *Boswellia* Growth

All plants require the three basic nutrients known as NPK (nitrogen, phosphate and potassium) to grow well and maintain a healthy life. Both organic and chemical fertilizers are commercially available on the market. Organic fertilizer has a lower NPK ratio and does not have any chelated metallic (which is easily absorbed by the roots) elements compared to chemical fertilizer. Usually, organic fertilizers are more useful for better plant growth. However, potted plants sometimes face harsher conditions because they cannot obtain all the required nutrients from organic fertilizers. The reason is because irrigation sometimes washes away the organic material before plants can take it up. Therefore, chemical fertilizers are vital and necessary for healthy plant growth, especially during plant propagation times. For *Boswellia* plant growth, basic fertilizer with balanced NPK and chelated metal are sufficient. However, for various *Boswellia* species, some additional elements are needed for better plant growth. Some growers use seaweed-based fertilizer for better growth by some humid-tropical species such as *B. serrata*, *B. ovalifoliolata* and *B. dalzielii*.

Similarly, for *Boswellia* propagation and flowering, sufficient nutrients are necessary. Plants should be fed well throughout the year, and stopping their feeding is recommended during dormancy. Furthermore, to increase the vegetative growth of cuttings, the fertilizer should be enriched with nitrogen.

Cultivation Practices and Conservation of *Boswellia* Populations

Populations with a high level of genetic variation have been observed in different *Boswellia* species (Bekele, 2016). However, some *Boswellia* species such as *B. papyrifera* show a degraded size, spatial fragmentation and a lack of recruitment of new tree generations (Eshete, Sterck, & Bongers, 2011; Groenendijk, Eshete, Sterck, Zuidema, & Bongers, 2012; Tolera, Sass-Klaassen, Eshete, Bongers, & Sterck, 2013). Furthermore, in situ management and conservation sampling guarantee that most extant species are in safe hands and are preserved. However, for conservation of population diversity, genetically distinctive populations should be the focus, while the broader core of the population should be preserved, because populations can sometimes lose diversity due to genetic drift (Namkoong, 1988).

Pollination and Seed Formation

Some *Boswellia* species need more care and attention to yield flowers during cultivation. However, *Boswellia* flowering is species-specific and depends on the location of the plant (pots or in the ground). Some extra measures should be taken for *Boswellia* flowers outside of their natural environment (Eslamieh, 2017). For example, in some species such as *B. dioscoridis* and *B. ameero*, obtaining flowers is very difficult, while in some species such as *B. sacra* and *B. nana*, flowering and pollination are very easy to elicit with the help of pollinators such as bees, ants, etc., or also mechanically. All *Boswellia* species usually have perfect flowers, and one plant is sufficient to propagate for seed, but cross-pollination was found to be more useful than self-pollination. Furthermore, some species such as *B. nana* and *B. popoviana* cannot produce seeds through self-pollination (Eslamieh, 2017).

After 1–3 months of pollination and flowering, seed pods will be ready for harvest. However, this period varies from species to species. The seeds lose their colour after the ripening stage and change to light brown from light green, and after some time, the stalk will turn brown. In some species such as *B. sacra*, a single stalk produces many seed pods.

Germination Rate

Variation was observed in the germination rates of different *Boswellia* species and most can geminate easily. Some species have a high germination rate such as *B. nana*, while in contrast, *B. sacra* has one of the lowest germination rates (with the highest seed production). Similarly, the viability of *Boswellia* seeds also varies with the conditions (humidity, temperature and light), and most seeds can grow after 3 or more years. Various researchers have tried to increase the germination rate of *Boswellia sacra*. However, Eslamieh (2017) reported that no significant improvement was found to increase the germination rate of *B. sacra* from 10% under different techniques. Jason reported that the reason for a lower germination rate is empty pyrenes with no embryos, and genetic aberration plays more of role than sowing techniques. Furthermore, he cross-pollinated various *B. sacra* clones and observed better results (Eslamieh, 2017). The most prominent super *sacra* clone produced seeds with 40–60% germination rates following the cross-pollination of three generations.

Macropropagation

Clonal propagation technique plays a vital role in conservation and multiplication and in the large-scale production of an endangered forest tree species. This technique is most valuable for true-to-type plant production. Furthermore, clonal propagation using tissue culture technique provides a substitute for the vegetative practices used in the past. It has the potential to yield a higher multiplication of uniform genotypes in a short time and provide a substitute for the species that are difficult to grow (Gupta et al., 1993). There are many ways to propagate trees vegetatively, including *B. serrata*, an important medicinal and broadly useful species (Ghorpade, Chopra, & Nikam, 2010; Purohit, Tak, & Kukda, 1995; Sharma, 1983) of tree that grows in the hilly areas of India (Kapoor, 1990; Shah, Rathod, Suhagia, Pandya, & Parmar, 2008). Rooting *B. serrata* stem cuttings is very easy, inexpensive and successful for mass multiplication (Eslamieh, 2017).

Vegetative proliferation through stem cuttings has yielded positive results (Haile, Gebrehiwot, Lemenih, & Bongers, 2011). These proliferation methods are very useful techniques for producing high-quality propagules for various domestication programmes, genetic conservation and reforestation. Similarly, the perfect timing for propagule collection (Danthu et al., 2002; Welander, 1995) and the position and age of the parent stock (Husen & Pal, 2007; Opuni-Frimpong, Karnosky, Storer, & Cobbinah, 2008) are reported to have vital effects on these vegetative growth experiments. Some researchers have recommended basal long soak methods for treating these cuttings (Kroin, 1992, 2011). After auxin treatment, the cuttings were sown in root trainer (500 cc-conical cells) containing rooting media (sand and soil). One-third of the cut basal portion was inserted into different rooting media. Placing the cuttings in rooting medium is called 'sticking'. The medium was dampened with autoclaved water and the surface was made even. For *B. serrata*, the different auxins IBA (indole-3-butyric acid), IAA (indole-3-acetic acid) and NAA were used, and IBA showed a better result. Stem cuttings of *B. serrata* were treated with 500 ppm IBA for 15 min, which yielded a better result in combination with sand: soil (1:1) rooting media (Vijay, 2014). At increased IBA concentrations, there were decreased rooting percentages. When stem cuttings were treated with different concentration of IAA and NAA, the cuttings showed low rooting % and sprouting, and the cuttings survived for nearly 10 days (Vijay, 2014).

Similarly, the survival rates of the cuttings also depended on their timing. For example, high average survival rates were obtained from *B. papyrifera* cuttings when they were collected in May. However, the March collection yielded propagules that longer than one metre with greater than 50% survival rates. Different results were obtained from the work conducted by Haile et al. (2011), who reported that February and March are the best months for obtaining cuttings with higher survival rates. The lowest survival rates were obtained during the May collection. This finding might be due to the differences in the various agroecological locations of these populations, especially in the phenological calendar (Haile et al., 2011). Similarly, the highest average survival rate was obtained when the cuttings were collected in May under dormant conditions, while the lowest survival rate was observed in October during leaf shedding season. Similarly, Danthu et al. (2002) reported that the performance will be better when sampling during dormant ortet compared to active ortet. The reason might be the carbohydrate composition and accumulation in the cuttings or the high temperatures as reported by Leakey, Chapman, and Longman (1982) and Leakey and Coutts (1989).

Seedling Propagation in Nurseries

Information related to *Boswellia species* seedling management in nurseries is limited. However, as shown in various dryland species, seedlings can easily be grown in nurseries without problems. These seeds readily germinate, and seedlings grow vigorously on various soil types. However, as discussed earlier, various types of soil and different watering durations are recommended. On average, seedlings will need 4–5 months to reach a plantable size (Eslamieh, 2017).

Planting and Protecting Seedlings

At planting time, *Boswellia* seedlings should be healthy with sufficient vigour. Because the juvenile seedlings cannot grow well in harsh environments (arid and semiarid regions), the seedlings should be hardened off carefully at the time of planting (Eslamieh, 2017). After that, the *Boswellia* seedlings should be protected from grazing animals, fire and other pathogens. The construction of fencing and fire breaks will help to protect these seedlings. Similarly, various researchers have suggested using some insecticides to protect these seedlings (Eslamieh, 2017).

Root Cutting Propagation

Boswellia seedlings growing in nurseries were mostly reported to show poor success rates when grown in the field, even under protected conditions. However, vegetative propagation through root cuttings is predicted to give good results in field experiments for establishing some species, including for plantation forests (Eslamieh, 2017). Researchers have reported that similar to most deciduous plants, *B. papyrifera* can grow from branch cuttings (Haile et al., 2011). Moreover, macropropagation techniques are inexpensive, easy and rapid compared to other asexual methods such as budding, grafting and some micropropagation techniques (Lemenih & Kassa, 2011). The cutting material should be collected from a mature, carefully grown and healthy plant as reported earlier in *B. papyrifera*. Various factors such as the collection and planting times and the cutting length can play vital roles in the success rate of these cuttings. Mostly planting and collection are useful during the dry season when the donor's trees are under dormant conditions. For instance, in Tigray and Metema, the best months to collect and plant cuttings are February and March. Experiments have been initiated to explore options to minimize the heights and diameters of cuttings required for successful vegetative propagation (Lemenih & Kassa, 2011).

After being cut from the donor tree, these cuttings should be planted instantly to protect them from dehydration (Lemenih & Kassa, 2011). Furthermore, the tips of these cuttings must be covered with and should be treated with fungicide to avoid contamination and dehydration. Regular hoeing and weeding are required, and the cuttings should be protected from animals and termites. Some exogenous plant hormones such as auxins should be applied to branch cuttings to enhance and stimulate rooting. In *B. papyrifera*, a latex (containing indole acetic acid) obtained from *E. abyssinica* was used to initiate rooting as a replacement for auxin (Negussie, Aerts, Gebrehiwot, Prinsen, & Muys, 2009).

Root Sucker Propagation

Root suckers are the other way to produce *Boswellia* species. This technique also plays a vital role in *Boswellia* species propagation in natural environments (Eslamieh, 2017). Furthermore, propagation by this method employing the exposed roots of a mature *B. papyrifera* tree can initiate a root sprout, which can grow into a mature tree under protected conditions. The rate of natural propagation of *B. serrata* is too slow when older trees are propagated through root suckers. However, the root sucker technique has not been exploited commercially (Lemenih & Kassa, 2011).

Root Tuber Propagation

Most *Boswellia* species can develop into new plants by developing tubers and thickened roots, and these tubers and roots can be removed from the plant by using tuber propagation (Eslamieh, 2017). The critical part of this procedure is to ensure that the end parts of the cutting are potted in a well-drained potting mix and that a few inches of the top parts of the root or tuber are kept above the soil. Irrigation can play a vital role in getting oxygen to the roots, and it should be reported for all *Boswellia* species. Protection from the fungus and other pathogens involves constant moisture environments. Furthermore, in some *Boswellia* species such as *B. sacra*, it is possible to choose tubers with interesting forms and expose a larger portion of the tuber above the soil to develop an interesting caudex for staging. The exposed portion of the tuber will eventually have all the attributes of a seed-grown plant with exfoliating bark and a thickened base (Eslamieh, 2017).

Seedling and Rooted Cutting Requirements

When root systems have developed in the cuttings and seedlings, it is important to provide some fertilizers with a 5-10-10 NPK ratio and regular irrigation. Full-strength fertilizers are not recommended by growers because sometimes, too much fertilizer causes excessively rapid growth but thin stems, so the plant can stand without support. When the cutting has grown up to a few inches, then we can increase the concentration of fertilizers as the cutting grow larger (Eslamieh, 2017).

Almost all *Boswellia* species can be transferred from one place to another, and they are easy to re-establish. If the root ball is disturbed either accidentally or deliberately, most *Boswellia* species will lose their leaves almost completely. However, during the growing season, the plants will leaf out again within a couple of weeks. The resting period is the best time to transplant these species because the root ball can remain intact. Furthermore, cold and wet soil in combination is harmful to tropical succulents as well as to *Boswellia*. Transplanting should be performed when the soil is dry (Eslamieh, 2017).

Micropropagation

In Vitro Micropropagation of Boswellia

In vitro techniques are used for the conservation of important plants, especially those that are difficult to propagate by conventional means. The micropropagation of various *Boswellia* species has already been reported by various researchers (Chandrasekhar, Hussain, & Jayanand, 2005; Purohit et al., 1995). The tissue

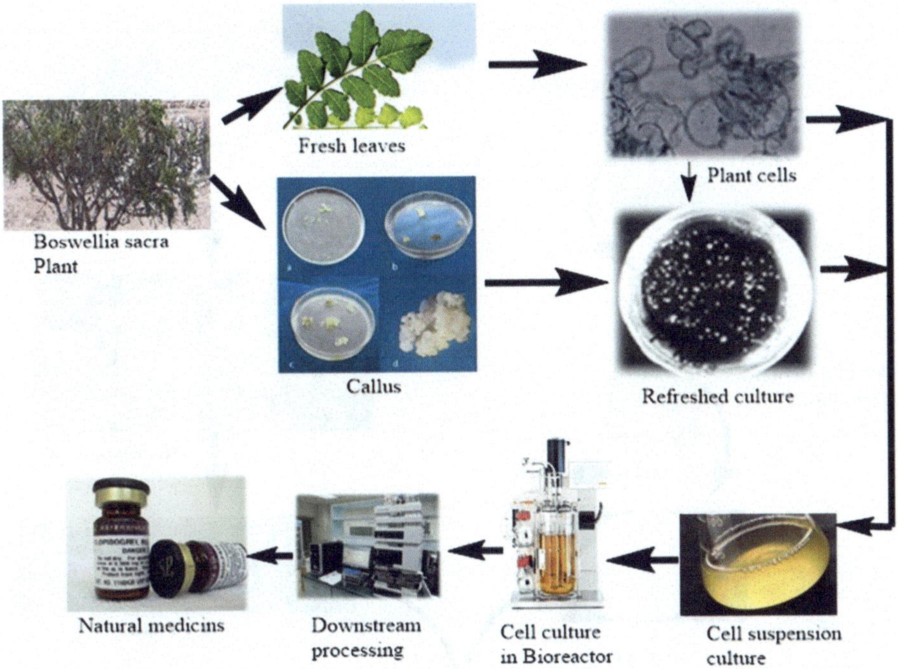

Fig. 5.1 Overview of drug development from *Boswellia sacra* via tissue culture and callus fermentation

culture technique is one of the efficient tools that can be used to achieve the conservation goals with respect to rare, endangered and threatened plants (Ahire, Ghane, Lokhande, Suprasanna, & Nikam, 2011). The use of micropropagation, particularly *Boswellia sacra* micropropagation and fermentation for drug development, is presented below in Fig. 5.1.

Axillary Bud Break and Axillary Shoot Proliferation

Approximately 0.1% and 0.2% of $HgCl_2$ was used for *B. serrata* surface sterilization at the nodal segment containing axillary buds, for 5–10 min (Vijay, 2014). Reportedly, 0.2% $HgCl_2$ can increase the contaminant-free percentage of explants relative to 0.1%. However, strong treatment can cause injury in the explant and lead to the death of plant. To get rid of browning effects and prevent the death of explants, the adsorbent PVP (polyvinylpyrrolidone), 100 mg/l antioxidant (50 mg/l ascorbic acid, 30 mg/l citric acid) or quick sub-culturing were performed to overcome these problems (Vijay, 2014). In Fig. 5.2, the general overview about callus induction is presented.

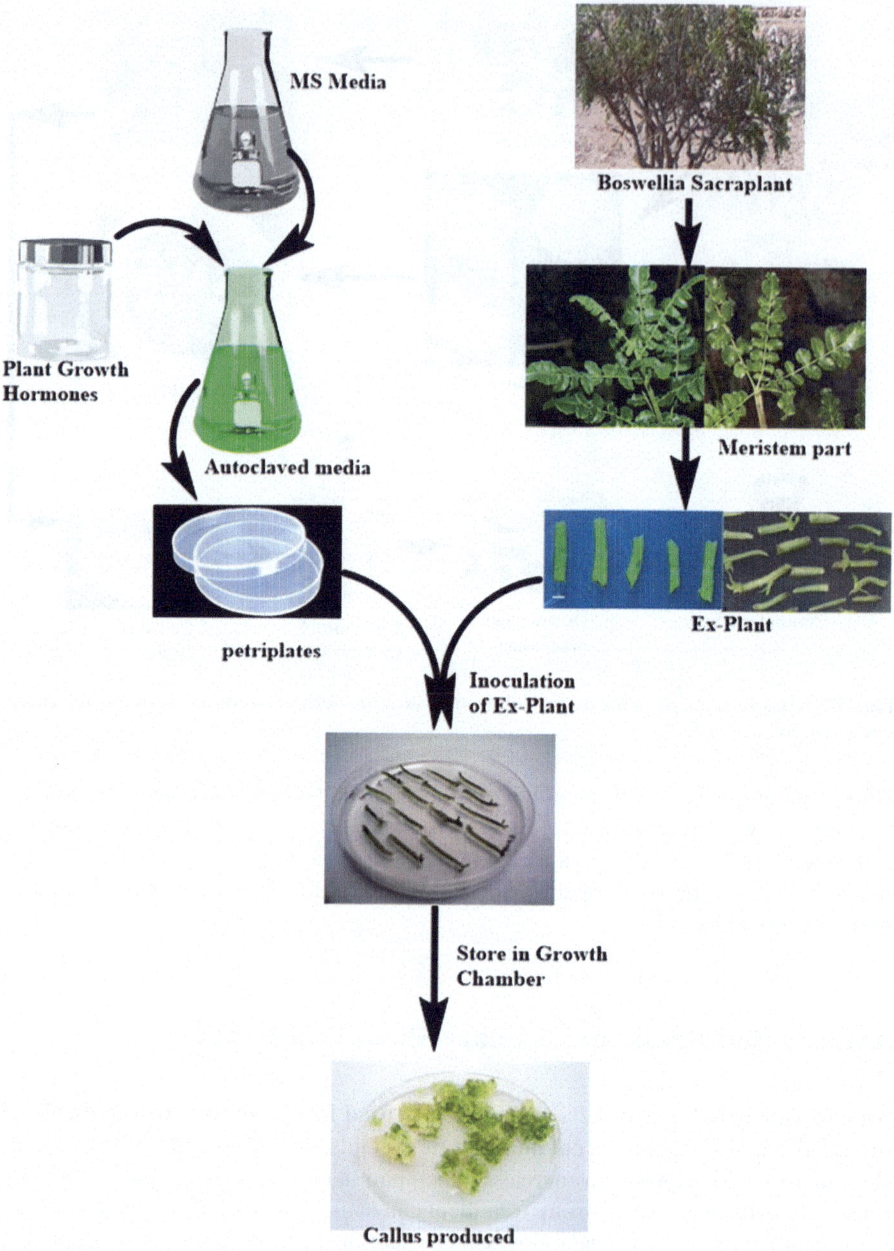

Fig. 5.2 General overview of *Boswellia sacra* callus culture induction

Effects of Seasonal Changes on Explants

As in micropropagation, the explants collected during different seasons of the year also showed variations in the percentages of axillary bud break responses. The nodal segments collected from September to November showed maximum bud break responses (Vijay, 2014). Nodal segments collected during other months produced reduced axillary bud break responses either due to the contamination of the explants, phenolic exudation or due to the presence of dormant meristems during the cold season. Explants collected during the rainy season were more susceptible to contamination. Thus, the results showed that February to April was the appropriate time for nodal explant collection and the axillary bud break response (Vijay, 2014).

Effects of Phytohormones

Two types of cytokinins, TDZ and BAP, were investigated individually using different concentrations for auxiliary bud proliferation (Vijay, 2014). The cytokinin presence strongly affected the bud induction on free cytokine media. In *B. serrata,* medium supplemented with BAP (8.88 μM) and additives (100 mg/l PVP +30 mg/l citric acid +50 mg/l ascorbic acid) showed the highest results (Vijay, 2014). The concentration of BAP medium influenced the response percentage. Furthermore, the bud break response was decreased at low concentrations while it increased with increased BAP concentration. Notably, the TDZ-supplemented MS medium did not show an increase in the axillary bud break response, and the maximum observed response was 20%. Furthermore, at higher concentrations, it also showed a decrease in the bud break response in *B. serrata* (Vijay, 2014). In Figs. 5.3 and 5.4 below, the callus was initiated using different plant hormones.

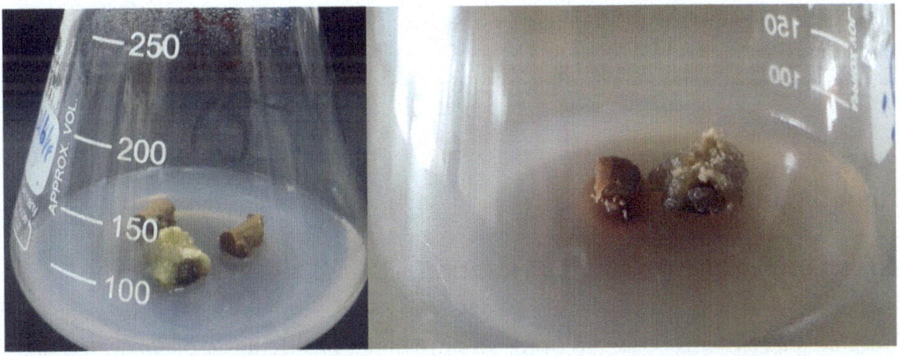

Fig. 5.3 Callus initiation in *Boswellia sacra* using different plant hormones. (Courtesy Muhammad Numan and Dr. Tapan Kumar Mohanta)

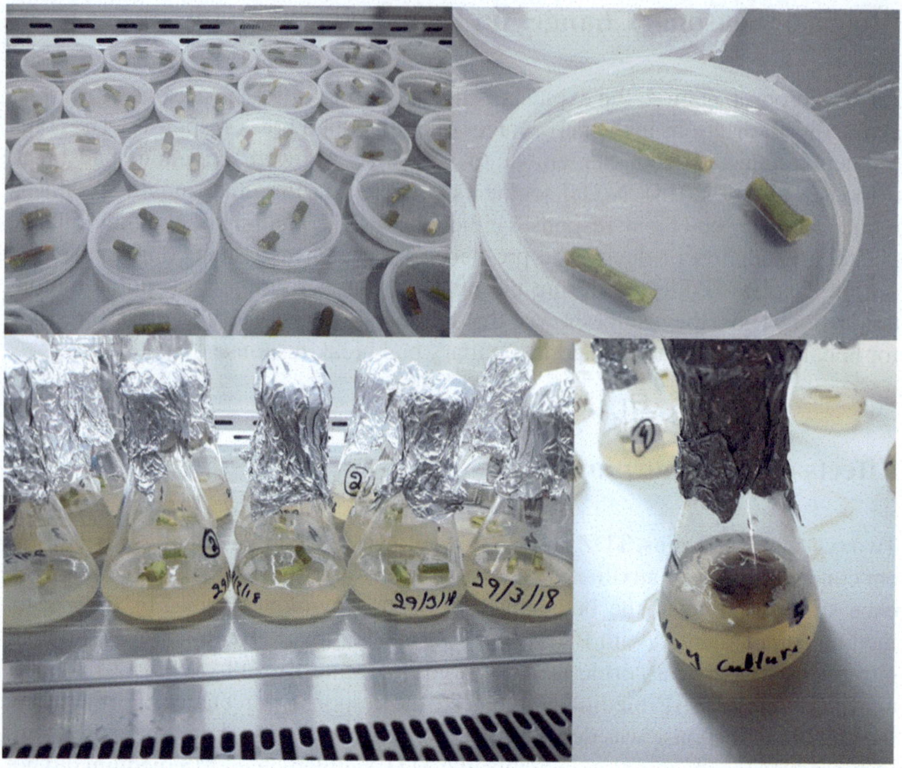

Fig. 5.4 Explant inoculation in media for callus induction. In the right corner, a flask containing a callus can be seen

Effect of Combined Auxin-Cytokinin Interaction

Compared to cytokine-only treatment, an increase was reported when cytokine was used in combination with small doses of auxin (NAA), and induction was observed from the *B. serrata* nodal segment (Vijay, 2014). Similarly, an increase was observed with an increase in the auxin concentration.

Shoot Multiplication

Effect of Plant Growth Regulators

Limited success was observed in the in vitro shoot proliferation of *Boswellia* species. It was reported that semisolid and liquid MS medium supplemented with both cytokinin and NAA did not show any positive effect on proliferated shoot multiplication, and very limited success was achieved (Vijay, 2014). After induction, the leaves turned brown and were found to have fallen on the medium, leading to stem necrosis and the death of the shoot. Furthermore, a number of hormonal combinations were checked on *B. serrata*. However, all attempts were unsuccessful except NAA and BAP-supplemented medium, in which some in vitro shoots survived for a few months.

All attempts at in vitro shoot multiplication, such as the use of different sources of energy (carbohydrates) in MS medium; the addition of different types of amino acids such as cystine, proline, L-arginine and L-asparagine to MS medium; the addition of antioxidants, silver nitrate and adenine sulphate; and the use of different sets of physical conditions (light and temperature), could not induce in vitro shoot multiplication, which evidently means that the culture has a different set of physico-chemical requirements (Vijay, 2014).

References

Ahire, M. L., Ghane, S. G., Lokhande, V. H., Suprasanna, P., & Nikam, T. D. (2011). Micropropagation of Uraria picta through adventitious bud regeneration and antimicrobial activity of callus. *In Vitro Cellular & Developmental Biology-Plant, 47*(4), 488.

Bekele, A. A. (2016). *Conservation genetics of the frankincense tree*. Wageningen, The Netherlands: Wageningen University.

Chandrasekhar, T., Hussain, T. M., & Jayanand, B. (2005). In vitro micropropagation of Boswellia ovalifoliolata. *Zeitschrift für Naturforschung C, 60*(5–6), 505–507.

Danthu, P., Soloviev, P., Gaye, A., Sarr, A., Seck, M., & Thomas, I. (2002). Vegetative propagation of some West African Ficus species by cuttings. *Agroforestry Systems, 55*(1), 57–63.

Eshete, A., Sterck, F., & Bongers, F. (2011). Diversity and production of Ethiopian dry woodlands explained by climate-and soil-stress gradients. *Forest Ecology and Management, 261*(9), 1499–1509.

Eslamieh, J. (2017). *The genus Boswellia: preservation through horticulture* (p. 80527). Fort Collins, CO: A Book's Mind.

Ghorpade, R. P., Chopra, A., & Nikam, T. D. (2010). In vitro zygotic embryo germination and propagation of an endangered *Boswellia serrata* Roxb., a source of boswellic acid. *Physiology and Molecular Biology of Plants, 16*(2), 159–165.

Groenendijk, P., Eshete, A., Sterck, F. J., Zuidema, P. A., & Bongers, F. (2012). Limitations to sustainable frankincense production: Blocked regeneration, high adult mortality and declining populations. *Journal of Applied Ecology, 49*(1), 164–173.

Gupta, P. K., Pullman, G., Timmis, R., Kreitinger, M., Carlson, W. C., Grob, J., & Welty, E. (1993). Forestry in the 21st Century. *Nature Biotechnology, 11*(4), 454.

Haile, G., Gebrehiwot, K., Lemenih, M., & Bongers, F. (2011). Time of collection and cutting sizes affect vegetative propagation of *Boswellia papyrifera* (Del.) Hochst through leafless branch cuttings. *Journal of Arid Environments, 75*(9), 873–877.

Husen, A., & Pal, M. (2007). Metabolic changes during adventitious root primordium development in *Tectona grandis* Linn. f.(teak) cuttings as affected by age of donor plants and auxin (IBA and NAA) treatment. *New Forests, 33*(3), 309–323.

Kapoor, L. (1990). *CRC handbook of Ayurvedic plants*. CRC Press, Boca Raton, FL, USA.

Kroin, J. (1992). Advances using indole-3-butyric acid (IBA) dissolved in water for-rooting cuttings, transplanting, and grafting. In *Combined proceedings-International Plant Propagators Society* (pp. 489–489). UNIV WASHINGTON-INT PLANT PROPAGATION SOC.

Kroin, J. (2011). How to improve cuttings propagation using water-based indole-3-butyric acid rooting solutions©. In *Combined proceedings International Plant Propagators Society*, Bellefonte, PA, USA.

Leakey, R., Chapman, V., & Longman, K. (1982). Physiological studies for tropical tree improvement and conservation. Factors affecting root initiation in cuttings of Triplochiton scleroxylon K. Schum. *Forest Ecology and Management, 4*(1), 53–66.

Leakey, R., & Coutts, M. (1989). The dynamics of rooting in Triplochiton scleroxylon cuttings: Their relation to leaf area, node position, dry weight accumulation, leaf water potential and carbohydrate composition. *Tree Physiology, 5*(1), 135–146.

Lemenih, M., and Kassa, H. (2011). *Management guide for sustainable production of frankincense: a manual for extension workers and companies managing dry forests for resin production and marketing*. CIFOR, Bogor, Indonesia.

Namkoong, G. (1988). Sampling for germplasm collections. *HortScience (USA), 23*, 79–81.

Negussie, A., Aerts, R., Gebrehiwot, K., Prinsen, E., & Muys, B. (2009). Euphorbia abyssinica latex promotes rooting of Boswellia cuttings. *New Forests, 37*(1), 35.

Opuni-Frimpong, E., Karnosky, D., Storer, A., & Cobbinah, J. (2008). Key roles of leaves, stock-plant age, and auxin concentration in vegetative propagation of two African mahoganies: *Khaya anthotheca* Welw. and Khaya ivorensis A. Chev. *New Forests, 36*(2), 115–123.

Purohit, S., Tak, K., & Kukda, G. (1995). In vitro propagation of *Boswellia serrata* Roxb. *Biologia Plantarum, 37*(3), 335–340.

Shah, S. A., Rathod, I. S., Suhagia, B. N., Pandya, S. S., & Parmar, V. K. (2008). A simple high-performance liquid chromatographic method for the estimation of boswellic acids from the market formulations containing *Boswellia serrata* extract. *Journal of Chromatographic Science, 46*(8), 735–738.

Sharma, S. (1983). Census of rare and endemic flora of south-east Rajasthan. In S. K. Jain & R. R. Rao (Eds.), *Assessment of threatened plants of India: proceedings of the seminar held at Dehra Dun, 14–17 September 1981* (p. c1983). Howrath, India: Botanical Survey of India, Department of Environment.

Tolera, M., Sass-Klaassen, U., Eshete, A., Bongers, F., & Sterck, F. J. (2013). Frankincense tree recruitment failed over the past half century. *Forest Ecology and Management, 304*, 65–72.

Vijay, N. (2014). Studies on in vitro cloning of mature Tree of capparis decidua and *Boswellia serrata*. Forest Research Institute University, Dheradun, India.

Welander, M. (1995). Influence of environment, fertilizer and genotype on shoot morphology and subsequent rooting of birch cuttings. *Tree Physiology, 15*(1), 11–18.

Chapter 6
Genetic Diversity and Differentiation Among Species and Populations of *Boswellia*

Genetic Diversity in Plants: Species vs Population

Genetic diversity refers to the level of variation in the genetic information within and/or among the populations of a species or a total community of species (Govindaraj, Vetriventhan, & Srinivasan, 2015; Nybom, 2004). Biodiversity is a broader term to express the concentration of biological diversity in relation to the number, variety and variation in living organisms in specific ecosystems. The population-based genetic diversity is essential to the existence of specific species (Table 6.1). This approach helps to elucidate the basis of phenotypic differences due to genetic manipulations (Bell & Collins, 2008), which play a vital role in the adaptation potential of populations (Hoffmann & Sgrò, 2011). In addition, genetic diversity shows a neutral level of differences in populations, suggesting a drift in breeding and genetics (Blows & Hoffmann, 2005). This diversity is directly proportional to the phenotypic variation of the population, to enhance its adaptability and evolution (Lanfear, Calcott, Kainer, Mayer, & Stamatakis, 2014).

The conservation and preservation of a species and population confronted with extreme risk of extinction requires multiple steps towards understanding and prioritizing genetic diversity. Within conservation strategies, categorical deliberations regarding genetic and evolutionary steps are essential for the sustainable use of resources for current and future generations (Laikre, Schwartz, Waples, & Ryman, 2010; Ponce-Reyes, Nicholson, Baxter, Fuller, & Possingham, 2013). Understanding the level of genetic variability of a plant species and the spatial distribution between populations is vital for recognizing the priorities for future conservation strategies (Edwards, Parchman, & Weekley, 2011; Moran, 2002; Verde et al., 2013). The genetic assortment of plant populations, among other influences, is directed by genetic and demographic practices, for example, the genetic drift, population size, gene dispersal of the species and the breeding system (Booy, Hendriks, Smulders, Van Groenendael, & Vosman, 2000; Charlesworth, 2009) and their relations to environmental factors (Lawton-Rauh, 2008; Ratnam et al., 2014).

A. Al-Harrasi et al., *Biology of Genus Boswellia*,
https://doi.org/10.1007/978-3-030-16725-7_6

Table 6.1 Glossary of critical terms in genetic diversity

Terms	Meaning
Gene flow	Transfer of genetic material from one population to another, also referred to as migration
Genetic paradox	A dilemma in invasion biology: how do newly founded populations overcome low genetic diversity and an expected low evolutionary potential, which is typically associated with extinction risk, to become established outside of their native range?
Hybridization	The breeding of individuals from genetically distinct populations (either within or between species), resulting in genotypes with novel combinations of alleles
Inbreeding depression	The reduction of fitness caused by mating between relatives
Vector	The physical conveyance responsible for moving introduced propagules from source to recipient environments
Allelic diversity	The number of different alleles per locus in a population. Allelic richness is a measure of allelic diversity as standardized by sample size
FST	A measure of genetic differentiation describing the proportion of genetic variation among populations relative to the variation within populations. The FST ranges from zero (no differentiation) to one (complete differentiation)
Genetic differentiation	A measure of how genetic diversity is partitioned among populations within a species
Genetic drift	Random changes in allele frequencies between generations due to chance alone
Genetic structure	Non-random distribution of genetic diversity that can form over space or time
Heterozygosity	Commonly measured as observed or expected heterozygosity. Observed heterozygosity (HO) is the average observed heterozygosity of individuals at the population level. The HE is the expected heterozygosity of individuals within populations under the assumption of Hardy-Weinberg equilibrium
Factors	Any variable describing sites or communities

Adapted from Banks et al. (2013) and This et al. (2006)

Molecular Markers for Analysing Genetic Diversity

Evaluating the genetic diversity of plant species depends on the use of diverse genetic markers (Lamy, Laroche, David, Massol, & Jarne, 2017; Moran, 2002; Ouborg, Angeloni, & Vergeer, 2010). Many recent genetic markers, such as amplified fragment length polymorphisms (AFLP), simple sequences repeat (SSR), random amplified polymorphic DNA (RAPD) and biochemical markers (SDS-PAGE), among other applications, are usually implied to the evaluation of a population's genetic diversity (Booy et al., 2000; Coppi, Cecchi, Selvi, & Raffaelli, 2010; González-Martínez, Krutovsky, & Neale, 2006). Genetic markers and the protein profiling of plants have been effectively and widely used to define the evolutionary and taxonomic aspects of several crops (Kaiser et al., 2016; Nisar et al., 2007).

Molecular studies are used to evaluate the first nucleotide variation in the ITS region of nuclear DNA of multiple accessions and, at a finer scale, to analyse the amount and distribution of inter- and intrapopulation genetic diversity and differentiation using inter-simple sequence repeat markers (ISSRs). The latter is a PCR-based fingerprinting technique that involves the amplification of DNA segment between two inversely oriented and closely spaced microsatellite repeat regions (Bornet & Branchard, 2001; Pradeep Reddy, Sarla, & Siddiq, 2002), and it is widely used in plant populations genetics thanks to the reliability and reproducibility of the results (Rakoczy-Trojanowska & Bolibok, 2004). In particular, it has been applied to assess genetic variations and population differentiation in a number of endangered angiosperm or gymnosperm species from different areas and climates (PÉRez-Collazos & CatalÁN, 2007).

SSR markers are advantageous for understanding polymorphisms, abundance and codominant inheritance with high reproducibility (Edwards et al., 2011; Powell, Machray, & Provan, 1996). However, AFLP is the most effective genetic technique for evaluating genetic diversity for conservation purposes, for DNA fingerprinting, and a precise co-amplification of great numbers of restriction fragments and genetic distances among individuals and populations (de Mendonça et al., 2012; Lawton-Rauh, 2008). RAPD molecular markers have been used extensively to examine the genetic diversity of plant populations (Maki, 1999; Sreekumar & Renuka, 2006) and to recommend phylogenetic hypotheses in plants (Carter & Sytsma, 2001; Kjølner, Såstad, Taberlet, & Brochmann, 2004). The results obtained from the RAPD technique are reliable and widely accepted (Collins, Mill, & Möller, 2003). In addition, SDS-PAGE analysis is broadly used owing to its ease of use and validity at defining the genetic structure of plant germplasm because total storage proteins are mostly free of environmental flux (Hameed et al., 2009).

Genetic Diversity of *Boswellia* Species

Although performing genetic diversity analyses at the species level and within and among populations is important to understand many questions related to conservation, in *Boswellia*, only two species and their population genetic diversity have been broadly studied through various molecular marker analyses to date. These species and populations include *B. sacra* and *B. papyrifera*. In the following section, various studies on both species are discussed.

Genetic Diversity of *Boswellia sacra*

As mentioned in a previous chapter about the location of *B. sacra*, this species is distributed from the south-western regions of Dhofar, and it spreads to the north-eastern scattered edges of the Hasik region (Raffaelli, Tardelli, & Mosti, 2008). In Hasik, it grows over large, semi-desert areas, demonstrating that it is a keystone species (Raffaelli, Mosti, & Tardelli, 2003, 2006). *B. sacra* populations have shown clear signs of decline due to lower rejuvenation and anthropogenic stochastic influences (Haile, Gebrehiwot, Lemenih, & Bongers, 2011; Khan et al., 2016, 2017). Consequently, awareness of the ecology and distribution of the different populations in Oman is now improving (Raffaelli et al., 2006) (Fig. 6.1). Despite its rapidly declining status, low regeneration capacity, seed germination, lower annual growth rate and economic position (Addisalem, Bongers, Kassahun, & Smulders, 2016; Addisalem, Duminil, Wouters, Bongers, & Smulders, 2016; Teshome, Eshete, & Bongers, 2017), a few conservation efforts are underway, and most of them are not supported by genetic information. Genetically depauperate tree populations are more susceptible to anthropogenic and environmental conditions and are more prone to extinction than species with high genetic diversity (Huenneke, 1991; Fleishman et al., 2001).

In one of our unpublished recent studies, detailed SSR, AFLP, RAPD and SDS-PAGE analyses of 13 different populations of *B. sacra* were performed. Previously, Coppi et al. (2010; see preceding section) performed genetic diversity-related work on *B. sacra* using ITS and ISSR analyses with two primer sets. However, in the current study, we have expanded the number of populations as well as the marker used for molecular analysis to understand the current status of *B. sacra* and its populations. The aim of the present study was to evaluate the genetic diversity and genetic

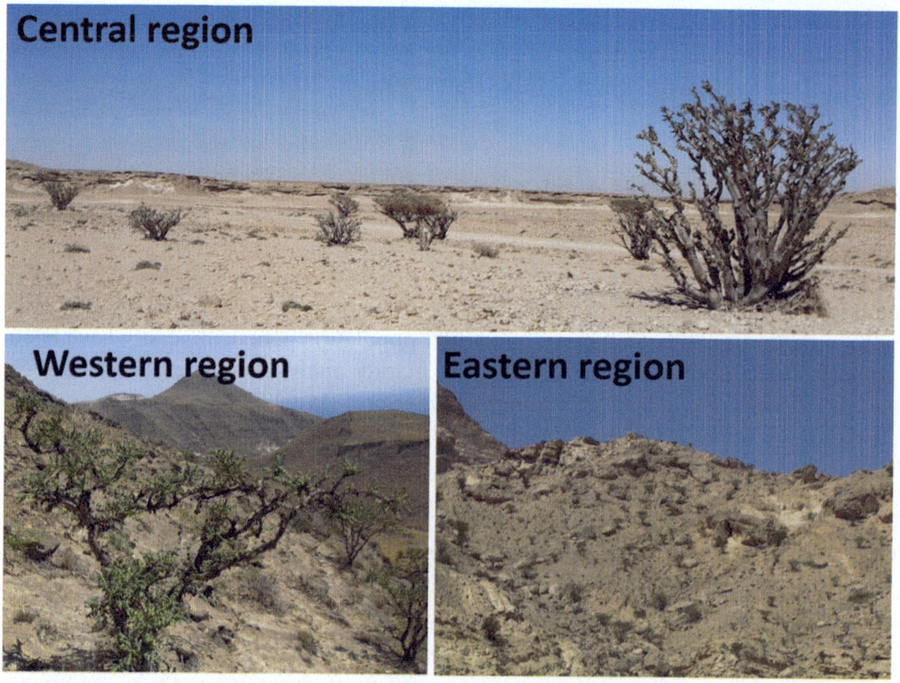

Fig. 6.1 Plant habitats and their distribution across three major areas of *B. sacra* populations in the Dhofar region

Table 6.2 List of evaluated *B. sacra* populations with geographic locations

NO.	Populations	Geographic coordinates	Elevations (m)
Central region			
P1	Dowkha (wild)	17 21.407'N 54′ 03.637′E	645
P2	Dowkha (cultivated)	17 20.273'N 54′ 04.580′E	663
Western region			
P4	Mughsil	17 20.260'N 54′ 20.260′E	600
P5	Adonab (wild)	16 52.573'N 53′ 44.516′E	181
P6	Adonab (cultivated)	16 57.893'N 53′ 52.705′E	97
	Sudh	16 57.225'N 53′ 54.839′E	57
Eastern region			
P7	Sadah -a	17 06.039'N 55′ 03.390′E	98
P8	Sadah -b	17 04.763'N 55′ 04.052′E	68
P9	Sadah -c	17 00.589'N 54′ 08.170′E	244
10	Hasik (lower)	17 31.237'N 55′ 12.923′E	215
P11	Hasik (upper)	17 32.102'N 55′ 13.320′E	254
P12	Kabout (lower)	17 44.988'N 55′ 21.143′E	474
P13	Kabout (upper)	17 47.821'N 55′ 24.559′E	31

differentiation of 13 populations (central, eastern and western regions; Table 6.2 and Fig. 6.1) of *B. sacra* growing wildly and in cultivated areas using AFLP, SSR, RAPD and SDS-PAGE markers (Table 6.3).

The results showed that in the case of the SSR markers, 88% of the bands were polymorphic and three markers produced the highest percentage (100%) of polymorphic bands. In addition, 8 AFLP markers and their analysis showed 1431 bands, out of which 1363 (95%) were polymorphic for the marker combinations *Eco*RI-AAG + MseI-CAT and *Eco*RI-AGT + *Mse*I-CAT. The authors also used 5 RAPD markers, which showed 34 (85%) polymorphic bands. In addition, to understand the population differentiation, protein polypeptides analysed through SDS-PAGE showed lower variations in 13 populations of *B. sacra* (Table 6.4).

In the interpopulation-based analysis, the F_{ST} value was high at 0.567 and the gene flow was 0.312 on the basis of the SSR markers analysis (Table 6.4). Mostly, a large genetic distance was observed between the eastern (Mughsil and Sadah) and western (Mughsil and Hasik) populations, which are geographically the most separated. Mughsil and Sadah were genetically the most isolated populations from one another (0.89) followed by Mughsil and Hasik (0.88) and Mughsil and Dowkha (0.81). Similar prospects were noted for AFLP. For RAPD molecular analysis, a large genetic distance was noted between the eastern (Sadah, Kabout and Hasik) and central region (Thumrait) populations, which are geographically the most isolated. Hasik and Thumrait were genetically the most isolated populations from one another (0.68) followed by Kabout and Tamrait (0.66). An AMOVA analysis showed that 69% (SSR), 72% (AFLP), 68% (RAPD) and 81% (SDS-PAGE) of the molecular variance were present within the populations and the lowest molecular variance among the populations. These five populations corresponded to diverse

Table 6.3 The primers sequences used for *B. sacra* populations

Primer pair	Sequences (5′-3′)	TB	PB	PPB%
SSR primers				
BP21R/BP21F	CTA TGT TCA TAC TTA GAA AAG TTG GTT (CTT)15 C	3	3	100
BP21R/BP21R	CAG AGT TAA TAA TAT AAG TAG CAG CAT (TTC)15TC	3	3	100
BP22R/BP22F	AGA ACC AGA CCT TCA AAT CAT (TCT)16TC	3	3	100
BP22R/BP22R	TAA AAC CAT TTT CAG CAA GGT T(TCT)16TC	3	3	100
BP23R/BP23F	TAA GAC CCC AAG AAA TTG AAT(TCT)20 TC	2	2	100
BP23R/BP23R	GCG AAT TTG CTC TGT AAT (TCT)20 TC	2	2	100
BP24R/BP24F	CAG TCT AAG TCC ACA AAC TCC CGG GGC GGG GCG GGG	1	0	0.00
BP24R/BP24R	TAT TTG TCA ACA GAT TGG GGC GGG GCG GGG CGG GG	1	0	0.00
Total		**9**	**8**	**88%**
AFLP primers				
(E-AAG/M-CAA)		179	164	92
(E-AAG/M-CAT)		230	193	84
(E-ACA/M-CAA)		154	152	99
(E-ACA/M-CAT)		200	195	98
(E-ACA/M-CGT)		180	179	99
(E-AGT/M-CAA)		150	146	97
(E-AGT/M-CAT)		164	164	100
(E-AGT/M-CGT)		174	170	98
Total		**1431**	**1363**	**95.80%**
RAPD primers				
OPA04	AAT CGG GCT G	7	6	85
CB9	GGT GAC GCA G	8	6	75
CB12	AGT CGA CGC C	8	7	87
CB13	ACG CAT CGG A	8	7	87
CB23	CTG GGC ACG A	9	8	88
Total		**40**	**34**	**85.00%**
SDS-PAGE (polypeptides)				
Total		**12**	**4**	**33%**

TB total bands, *PB* polymorphic bands, *PPB* percentage of polymorphic bands, *E* EcoliR, *M* Msel

Table 6.4 Total mean band frequencies, estimated allele frequencies and estimated heterozygosity and other analyses for all populations

Total	N	Na	Ne	I	He	uHe	Ho	F
SSR	2.769	1.596	1.464	0.355	0.241	0.298	0.304	−0.255
AFLP	2.769	2.6	1.82	0.893	0.312	0.723	0.569	
RAPD	2.769	1.194	1.221	0.188	0.127	0.156		
SDS-PAGE	2.769	1.167	1.148	0.122	0.083	0.102		

N no. of samples, *Na* no. of different alleles, *Ne* no. of effective alleles, *I* Shannon's Information Index, *He* expected heterozygosity, *uHe* unbiased expected heterozygosity, *Ho* observed heterozygosity, *F* fixation index = (He − Ho)/He = 1 − (Ho/He)

geographical regions and to different environmental conditions, i.e. three populations (Thumrait, Dowkha-a and Dowkha-b) are from central regions, and one of each population was from Mughsil and Sadah in the western and eastern regions, respectively. The Mughsil population has maximum geographic and genetic distances from the rest of the populations (Fig. 6.2).

Population Genetic Diversity and Conservation Issues in *B. sacra*

Among other approaches, conservation genetics offers an understanding of the gene flow and dynamics in a population to counteract species extinction and support population restoration (Barnes & Turner, 2016; Pinares, González-Astorga, Vovides, Lazcano, & Vendrame, 2009). It is emphasized for technically sound management and the conservation of genetic assets. In the current study, the overall results showed different allele frequencies in populations of *B. sacra* according to the SSR (0.127–1.194), RAPD (0.312–1.82) and AFLP (0.083–1.148). The expected heterozygosity levels (H_E) for SSR, AFLP, RAPD and SDS-PAGE were 0.241, 0.312, 0.127 and 0.083, respectively, among 13 populations, which is comparatively higher than the earlier report by Coppi et al. (2010). The previous study showed only one type of haplotype-based measurement, in which the reported expected heterozygosity (H_E) for ISSR markers ranged between 0.11 and 0.18 for different populations of *B. sacra*. The total heterozygosity (H_T 0.413) according to the SSR markers was higher in *B. sacra* compared to earlier studies using ISSR markers (H_T 0.22). However, when pairwise F_{ST} values were assessed, there was comparatively higher gene flow among Dowkha (cultivated), Dowkha (wild), Sadah-b and Kabout-lower samples. By contrast, a lower gene flow was observed for rest of the populations.

The populations were geographically separated into three regions, i.e. central regions with higher altitudinal ranges (600–645 m above sea level), eastern (31–474 m above sea level) and western coastal regions (57–181 m above sea level) with lower altitudinal ranges (31–474 m above sea level). For inter-population genetic differentiation, a large genetic distance was observed between the eastern (Mughsil and Sadah) and western region (Mughsil and

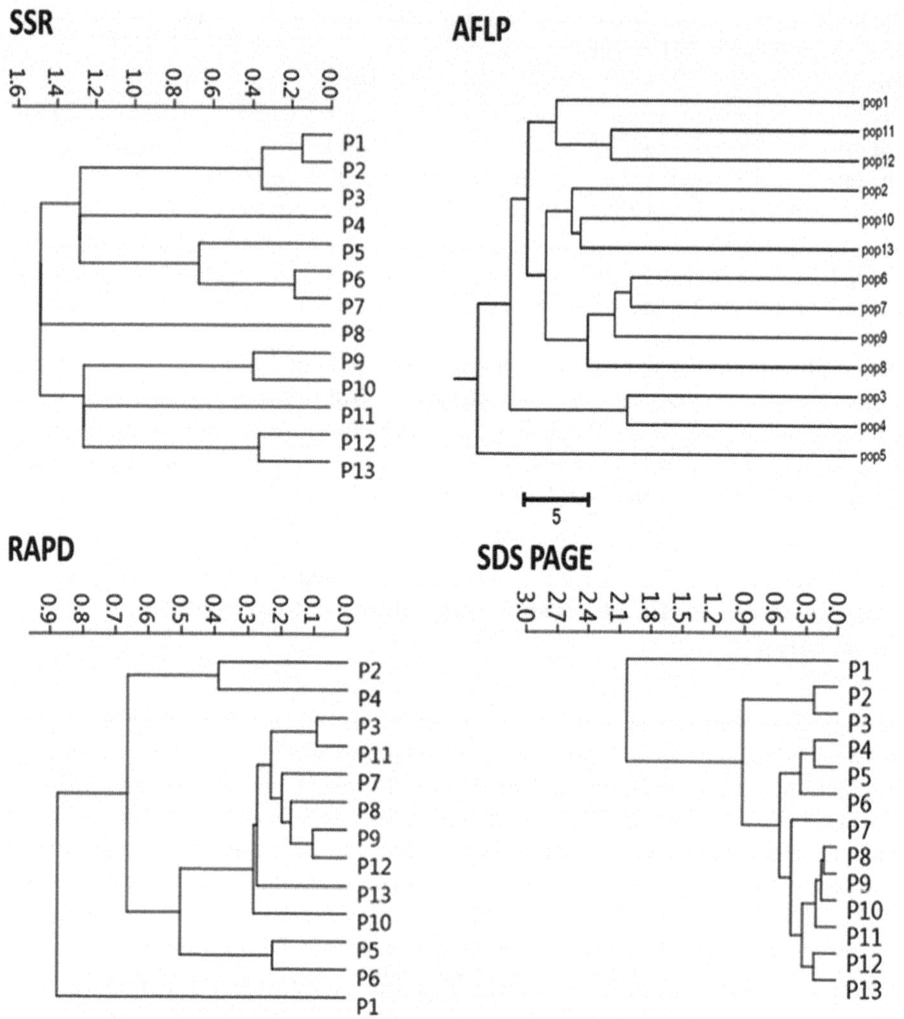

Fig. 6.2 Unweighted pair group method with arithmetic (UPGMA) cluster analysis of the *B. sacra* populations based on genetic distances using SSR, AFLP, RAPD and SDS-PAGE analyses. This analysis was performed using PAST v.3.01

Hasik) populations, which are geographically most detached from one another. Surprisingly, a lower genetic diversity among populations of the central region was observed, and one of these populations (Dowkha) was from the national park, where the frankincense trees are well maintained. However, the isolation reflected by the distance model showed the correlation between genetic and geographic distances. This relationship may be attributed to the fact that the species is outcrossed and wind pollinated and may engage in seed dispersal through water in eastern and western coastal populations (Coppi et al., 2010)

with respect to the commercial exploitation of the resin. Coppi et al. (2010) analysed 7 populations from 3 distinct geographic areas of Dhofar, which yielded 93 scorable bands through 4 ISSR primers.

Under the assumption that each fragment represents a genetic locus, it was possible to estimate the allelic frequency and therefore to proceed further towards estimating classical population parameters. Large populations with cross-pollinated species usually have broad genetic diversity. In long-lived cross-pollinated plant species, the bulk of the genetic diversity resides within populations (Hamrick, 2004). Populations located away from one another will be genetically differentiated because of limited gene flow from the movement of individuals, or their pollen or seeds, over longer spaces (Schaal, Hayworth, Olsen, Rauscher, & Smith, 1998). Geographic distance alone may not describe all genetic differentiation since variable shapes may also be squeezed by environmental variables that are vital for the survival or growth of the species (Parmentier et al., 2013). Variations in the population demography as a consequence of a prompt drop in the population size may result in a considerable loss of genetic diversity and may lead to rapid differentiation across populations as a consequence of genetic drift if it lasts for more than a few generations (Frankham, 1995).

Despite the low genetic differentiation within populations, all the markers isolated the Dowkha population from the others. Dowkha is in the central region relative to where the *B. sacra* populations started. Apart from SSR marker data, a lower heterozygosity, which was not unexpected, was observed in the remaining molecular markers. The populations with higher genetic diversity such as Dowkha (wild) and Dowkha (cultivated) seem to be fragments of a once large population. Therefore, it could be suggested that the wild Dowkha population might also be considered as a part of the national park (Dowkha (cultivated)) for in situ and ex situ conservation efforts.

In addition, conservation efforts are hampered by a lack of uniform seed germination for maintaining the tree population. For example, in *B. papyrifera*, authorities germinated 15,000 seeds annually for regeneration, but the success rate was only 1–5% (Rijkers, Ogbazghi, Wessel, & Bongers, 2006). A similar conclusion was also drawn by Eslamieh (2017) about *B. sacra*, which possesses low flower and fruit production, reduced natural regeneration and non-viable seeds. In addition, tree cutting, animal grazing and leaf collection for fodder purposes are also some of the causes of *B. sacra* degradation (Raffaelli et al., 2008). There is an urgent need to control fungal infections and pest attacks by monitoring the oldest frankincense trees (Raffaelli et al., 2008). Due to non-uniform genetic variation in the frankincense trees of the Dhofar regions, the eastern region may be considered for ex situ conservation because there is higher genetic diversity (Coppi et al., 2010). Ex situ conservation measures may include seed collection and maintenance, seed propagation tissue culture propagation and plant breeding methods. Tissue culture may be used for plant regeneration, because it has also been used successfully for *B. sacra* from Oman using Murashige and Skoog (MS) with 2,4-dichlorophenoxyacetic acid (2,4-D) and indole acetic acid (IAA) (Quazi et al., 2014).

Genetic Diversity of *B. papyrifera*

Various processes, for example, gene flow and different stages of local selection, were used to determine the spatial genetic structure (Addisalem, Bongers, et al., 2016). In plant species, various factors such as seed- and pollen-mediated gene dispersal affected the genetic diversity within populations. The gene flow through both pollen and seeds has very important implications for the maintenance of genetic diversity (Moran & Clark, 2011; Peakall, Ruibal, & Lindenmayer, 2003). Furthermore, gene flow develops the genetic structure even in large continuous populations where the gene flow is low (Moran & Clark, 2012; Segelbacher et al., 2010; Vekemans & Hardy, 2004). Furthermore, in both annuals and perennials, the breeding system and population density of species are vital determinants of the intra-population genetic structure (Duminil, Hardy, & Petit, 2009). Similarly, low gene flow and the above-mentioned factors affect the genetically related individuals in combination and forms fine-scale spatial genetic structure (FSGS) (Vekemans & Hardy, 2004). Due to genetic drift and biparental inbreeding, the structure is protected if seed dispersal and pollen movement are limited (King & Zalucki, 2012; Moran & Clark, 2012).

However, broad dispersal and broad dispersal with movement would weaken FSGS over generations. Furthermore, gene dispersal within populations of species is very important to understand during the design of a conservation strategy that will be more important for genetic diversity maintenance (Addisalem, Bongers, et al., 2016). Previously, (Addisalem, Esselink, Bongers, & Smulders, 2015) used approximately six microsatellite markers to detect the genetic diversity and FSGS of two *B. papyrifera* populations. The genetic diversity was found both in terms of allele numbers and heterozygosity. A low fixation index, low Sp value and high degree of heterozygosity may indicate extensive intrapopulation gene dispersal. The Sp values detected for *B. papyrifera* were found to be almost similar to the average outcrossing (Addisalem, Duminil, et al., 2016). Similarly, the two taxonomically related species *B. serrata* and *B. ovalifoliolata* are outcrossing (Raju, Lakshmi, Ramana, & Chandra, 2012; Sunnichan, Mohan Ram, & Shivanna, 2005). These species also show a prezygotic self-incompatibility mechanism (Raju et al., 2012; Sunnichan et al., 2005). The heterozygosity results showed that outcrossing is the important mating system in *B. papyrifera* and may have a similar self-incompatibility system (Addisalem, Duminil, et al., 2016). Pollen dispersal involves gene dispersal in outcrossing species (Vekemans & Hardy, 2004). Moreover, the *B. papyrifera* gene dispersal detected within a population may be the result of the contribution from extensive pollen exposure. Cross-generation effects were strengthened by limiting the gene flow, and in *B. papyrifera*, it did not significantly change from adult to seedling (King & Zalucki, 2012; Moran & Clark, 2012; Smouse, Peakall, & Gonzales, 2008). In their study, Addisalem, Duminil, et al. (2016) reported that the adult population was higher in Kurmuk than in Guba-Arenja, but the Sp values were similar (0.014 and 0.011).

Similarly, various SSR markers were developed for the genetic diversity of *B. papyrifera* (Addisalem et al., 2015). Those researchers used approximately 46 SSR markers in the first set, which amplified up to 12 alleles in various individuals from almost 10 different locations in Ethiopia. Furthermore, they used the assessment, which is a technically easy screening step than can improve the marker efficiency, in outbreeding species. In *B. papyrifera*, the subset assessment method gave a significantly higher success compared to randomly chosen markers (Addisalem et al., 2015). Illumina paired-end DNA sequencing reads were used to develop the SSSSR markers in the young *B. papyrifera* leaves. Furthermore, the reads estimated a genome size of 705 Mb, which is close to 682 Mb for *B. serrata* (Addisalem et al., 2015). A similar size for *B. sacra* was also estimated recently (unpublished). The occurrences of multiple copies in a genome were analysed based on the sequence homology in k-mers (Koch, Platzer, & Downie, 2014). Furthermore, various retrotransposons, for example, gypsy and copia, were detected, and they are present in various numbers in the genomes (Addisalem et al., 2015). Mostly, the resources used for gene selection in this study were previously reported to be involved in the production of major resin production compounds in *B. papyrifera*, for example, diterpenes, triterpenes and nortriterpenes (Assefa et al., 2012; Basar, 2005; Bekana, Kebede, Assefa, & Kassa, 2014).

To facilitate population management and support the conservation of threatened species, this genetic information is one of the best tools to have (Edwards et al., 2011; Moran, 2002). Similarly, Addisalem et al. (2015) reported that the SSR markers developed for *B. papyrifera* can be useful for genetic diversity characterization, structure and processes of within populations, for example, seed dispersal and pollen distances, which may help to identify the conservation units of various species. Furthermore, a similar study on *B. papyrifera* population differentiation was performed in Ethiopia using SSR (Addisalem et al., 2015). Similarly, the polymorphism and cross-amplification of the SSR markers in *B. popoviana* and *B. pirrotae* show their importance in genetic studies of these species and may be helpful for other *Boswellia* species as well (Addisalem et al., 2015).

B. papyrifera Intrapopulation Genetic Diversity

Previously, Addisalem, Bongers, et al. (2016) examined the population differentiation pattern and genetic diversity of *B. papyrifera* in Ethiopia and found that its population significantly declined in size and was degraded and fragmented in various populations from different regions of Ethiopia (Eshete, Sterck, & Bongers, 2011; Tolera, Sass-Klaassen, Eshete, Bongers, & Sterck, 2013). A decrease in gene flow and a small population size may greatly reduce a population's genetic variability and drive strong spatial genetic structuring (Kramer, Ison, Ashley, & Howe, 2008; Schaal, et al., 1998). However, Addisalem, Bongers, et al. (2016) reported sufficient genetic variation and no inbreeding (HO = 0.669, HE = 0.681, FIS = 0.01) in a population of adult trees from various samples collected from different parts of

Ethiopia. Those researchers compared its heterozygosity with that of other tropical tree species such as *Terminalia/Combretum* and the *Acacia senegal* species co-occurring with *B. papyrifera* (Addisalem, Bongers, et al., 2016). It was reported that intraspecific genetic diversity is still present in mature *B. papyrifera* mature trees and may occur because of the long lifespan of these tress; the degraded ion and fragmentation events in Ethiopia are relatively recent events (Smulders et al., 2008). Furthermore, it is still unknown whether *B. papyrifera* can tolerate selfing, but its recruitment is one of the primary threats to its populations (Eshete et al., 2011; Tolera et al., 2013).

Similarly, Lemenih, Feleke, and Tadesse (2007) reported that continuous human interference was the cause of the degradation of these dry tropical forest ecosystems, and it increased by a high rate in the late nineteenth century. Furthermore, Tolera et al. (Tolera et al., 2013) observed the rings of trees and determined that the mature trees established themselves in the existing *B. papyrifera* forests from 1903 to 1955. Therefore, on the basis of the genetic data from the *B. papyrifera* population that was collected from various regions, these populations represent the diversity in their parents, which grew in continuous and large forests in the nineteenth century (Addisalem, Bongers, et al., 2016). The other important note was the discovery of the western region (Benishangul-Gumuz) population with its high density of saplings and seedlings, and natural regeneration occurs in this population (Addisalem, Bongers, et al., 2016).

Furthermore, (Addisalem, Duminil, et al., 2016) observed fewer agriculture activities, and the forests are mostly natural and undisturbed in this particular region. However, this condition is now changing, and even the livestock population in the Assosa region is still non-significant and low, especially in the *Boswellia* woodland regions. It was assumed that the high amount of rain and deep soil can cause microsites in seedling development (Alemu, Garedew, Eshetu, & Kassa, 2015; Eshete et al., 2011; Teshome, 2013).

Population Differentiation in *B. papyrifera*

Addisalem, Bongers, et al. (2016) reported that the average FST from 5 markers was 0.0109 ($P < 0.001$) among all 12 populations. However, the FST was significantly decreased after the exclusion of two populations (AF-BIR and JRG) (Addisalem, Bongers, et al., 2016). The isolation of the population, the geographical distance and environment adaptations may cause the highest FSTs in these populations. AF-BIR is geographically isolated, and JRG is the most distant from other populations because of the mountains. Furthermore, sampling is very important because young leaves are vulnerable under irregular and unpredictable rainfall. Therefore, another approach will be DNA extraction from bark tissue, but it is very difficult to extract good-quality DNA because of the exudate in the bark of *Boswellia* species (Novaes, Rodrigues, & Lovato, 2009).

Pollen dispersal by insects plays important roles in the isolation-by-distances pattern of genetic differentiation among different populations of *Boswellia* species. Moreover, occasional chances from long distance dispersal were also observed in regions where the species were distributed along riverbanks. However, the contribution of water to seed dispersal was found to be lower in *B. papyrifera* compared to various riparian species (Addisalem, Bongers, et al., 2016).

Similarly, Addisalem, Bongers, et al. (2016) also reported a clear isolation-by-distance (IBD) pattern, which showed that approximately 44% of the variation in the populations that support this *B. papyrifera* population are spatially structured because of their geographical isolation. However, the genetic differentiation may be due to natural selection and genetic drift (Duminil et al., 2013; Ledig & Fryer, 1972; Slatkin, 1987). Similarly, ten populations were resolved by the individual-based structure into four genetically differentiated clusters that correlated with the geographical locations of the sampled populations. These clusters, northern (N), western (W), north-eastern (NE) and north-western (NW), also link to different environmental conditions such as the soil conditions and precipitation (Addisalem, Bongers, et al., 2016). Hence, the genetic differentiation in the population that was detected in *B. papyrifera* may differentiate populations that are also adapted to local environmental conditions caused by precipitation, temperature, altitude and soil conditions or combinations of these factors. This hypothesis must be confirmed through further studies such as common garden experiments or provenance studies.

References

Addisalem, A., Bongers, F., Kassahun, T., & Smulders, M. (2016). Genetic diversity and differentiation of the frankincense tree (*Boswellia papyrifera* (Del.) Hochst) across Ethiopia and implications for its conservation. *Forest Ecology and Management, 360*, 253–260.

Addisalem, A., Duminil, J., Wouters, D., Bongers, F., & Smulders, M. (2016). Fine-scale spatial genetic structure in the frankincense tree *Boswellia papyrifera* (Del.) Hochst. and implications for conservation. *Tree Genetics & Genomes, 12*, 86–96.

Addisalem, A., Esselink, G. D., Bongers, F., & Smulders, M. (2015). Genomic sequencing and microsatellite marker development for *Boswellia papyrifera*, an economically important but threatened tree native to dry tropical forests. *Aob Plants, 7*.

Alemu, B., Garedew, E., Eshetu, Z., & Kassa, H. (2015). Land use and land cover changes and associated driving forces in north western lowlands of Ethiopia. *International Research Journal of Agricultural Science and Soil Science, 5*, 28–44.

Assefa, M., Dekebo, H., Kassa, H., Habtu, A., Fitwi, G., & Redi-Abshiro, M. (2012). Biophysical and chemical investigations of frankincense of *Boswellia papyrifera* from North and Northwestern Ethiopia. *Journal of Chemical and Pharmaceutical Research, 4*.

Banks, S. C., Cary, G. J., Smith, A. L., Davies, I. D., Driscoll, D. A., Gill, A. M., & Peakall, R. (2013). How does ecological disturbance influence genetic diversity?. *Trends in ecology & evolution, 28*(11), 670–679.

Barnes, M. A., & Turner, C. R. (2016). The ecology of environmental DNA and implications for conservation genetics. *Conservation Genetics, 17*, 1–17. https://doi.org/10.1007/s10592-015-0775-4

Basar, S. (2005). Phytochemical investigations on Boswellia species. University of Hamburg, Germany.

Bekana, D., Kebede, T., Assefa, M., & Kassa, H. (2014). Comparative phytochemical analyses of resins of Boswellia species (B. papyrifera (Del.) Hochst., B. neglecta S. Moore, and B. rivae Engl.) from Northwestern, Southern, and Southeastern Ethiopia. *ISRN Analytical Chemistry, 2014*, 1–9.

Bell, G., & Collins, S. (2008). Adaptation, extinction and global change. *Evolutionary Applications, 1*, 3–16. https://doi.org/10.1111/j.1752-4571.2007.00011.x

Blows, M. W., & Hoffmann, A. A. (2005). A reassessment of genetic limits to evolutionary change. *Ecology, 86*, 1371–1384. https://doi.org/10.1890/04-1209

Booy, G., Hendriks, R., Smulders, M., Van Groenendael, J., & Vosman, B. (2000). Genetic diversity and the survival of populations. *Plant Biology, 2*, 379–395.

Bornet, B., & Branchard, M. (2001). Nonanchored inter simple sequence repeat (ISSR) markers: Reproducible and specific tools for genome fingerprinting. *Plant Molecular Biology Reporter, 19*, 209–215. https://doi.org/10.1007/bf02772892

Carter, M. C., & Sytsma, M. D. (2001). Comparison of the genetic structure of North and South American populations of a clonal aquatic plant. *Biological Invasions, 3*, 113–118.

Charlesworth, B. (2009). Effective population size and patterns of molecular evolution and variation. *Nature Reviews Genetics, 10*, 195.

Collins, D., Mill, R. R., & Möller, M. (2003). Species separation of *Taxus baccata, T. canadensis*, and *T. cuspidata* (Taxaceae) and origins of their reputed hybrids inferred from RAPD and cpDNA data. *American Journal of Botany, 90*, 175–182.

Coppi, A., Cecchi, L., Selvi, F., & Raffaelli, M. (2010). The Frankincense tree (*Boswellia sacra*, Burseraceae) from Oman: ITS and ISSR analyses of genetic diversity and implications for conservation. *Genetic Resources and Crop Evolution, 57*, 1041–1052.

de Mendonça, P. C., Bertoni, B. W., Amui, S. F., Giuliatti, S., Corrêa, V. S. C., de Castro, F. S., & Pereira, A. M. S. (2012). Genetic diversity of *Stryphnodendron adstringens* (Mart.) Coville determined by AFLP molecular markers. *Biochemical Systematics and Ecology, 41*, 16–20.

Duminil, J., Brown, R. P., Ewédjè, E.-E. B., Mardulyn, P., Doucet, J.-L., & Hardy, O. J. (2013). Large-scale pattern of genetic differentiation within African rainforest trees: insights on the roles of ecological gradients and past climate changes on the evolution of Erythrophleum spp (Fabaceae). *BMC Evolutionary Biology, 13*, 195.

Duminil, J., Hardy, O. J., & Petit, R. J. (2009). Plant traits correlated with generation time directly affect inbreeding depression and mating system and indirectly genetic structure. *BMC Evolutionary Biology, 9*, 177.

Edwards, C. E., Parchman, T. L., & Weekley, C. W. (2011). Assembly, gene annotation and marker development using 454 floral transcriptome sequences in *Ziziphus celata* (Rhamnaceae), a highly endangered, Florida endemic plant. *DNA Research, 19*, 1–9.

Eshete, A., Sterck, F., & Bongers, F. (2011). Diversity and production of Ethiopian dry woodlands explained by climate-and soil-stress gradients. *Forest Ecology and Management, 261*, 1499–1509.

Eslamieh, J. (2017). *The genus Boswellia: preservation through horticulture* (p. 80527). Fort Collins, CO: A Book's Mind.

Fleishman, E., Launer, A. E., Switky, K. R., Yandell, U., Heywood, J., & Murphy, D. D. (2001). Rules and exceptions in conservation genetics: genetic assessment of the endangered plant Cordylanthus palmatus and its implications for management planning. *Biological Conservation, 98*(1), 45–53.

Frankham, R. (1995). Conservation genetics. *Annual Review of Genetics, 29*, 305–327. https://doi.org/10.1146/annurev.ge.29.120195.001513

González-Martínez, S. C., Krutovsky, K. V., & Neale, D. B. (2006). Forest-tree population genomics and adaptive evolution. *New Phytologist, 170*, 227–238.

Govindaraj, M., Vetriventhan, M., & Srinivasan, M. (2015). Importance of genetic diversity assessment in crop plants and its recent advances: an overview of its analytical perspectives. *Genetics Research International, 2015*, 431487.

Haile, G., Gebrehiwot, K., Lemenih, M., & Bongers, F. (2011). Time of collection and cutting sizes affect vegetative propagation of *Boswellia papyrifera* (Del.) Hochst through leafless branch cuttings. *Journal of Arid Environments, 75*, 873–877.

Hameed, A., Shah, T. M., Atta, B. M., Iqbal, N., Haq, M. A., & Ali, H. (2009). Comparative seed storage protein profiling of Kabuli chickpea genotypes. *Pakistan Journal of Botany, 41*, 703–710.

Hamrick, J. L. (2004). Response of forest trees to global environmental changes. *Forest Ecology and Management, 197*, 323–335. https://doi.org/10.1016/j.foreco.2004.05.023

Hoffmann, A. A., & Sgrò, C. M. (2011). Climate change and evolutionary adaptation. *Nature, 470*, 479. https://doi.org/10.1038/nature09670

Huenneke, L. F. (1991). Ecological implications of genetic variation in plant populations. Genetics and conservation of rare plants, 31–44.

Kaiser, R., Mageney, V., Schwefel, K., Vollmers, D., Krüger, A., & Horn, R. (2016). Genotyping of red and white fruited strawberry (Fragaria L.) accessions and hybrids based on microsatellite markers and on the genetic diversity in the allergen genes fra a 1 and fra a 3. *Genetic Resources and Crop Evolution, 63*, 1203–1217.

Khan, A. L., Al-Harrasi, A., Al-Rawahi, A., Al-Farsi, Z., Al-Mamari, A., Waqas, M., … Lee, I.-J. (2016). Endophytic Fungi from Frankincense Tree Improves Host Growth and Produces Extracellular Enzymes and Indole Acetic Acid. *PLoS One, 11*, e0158207. https://doi.org/10.1371/journal.pone.0158207

Khan, A. L., Al-Harrasi, A., Asaf, S., Park, C. E., Park, G.-S., Khan, A. R., … Shin, J.-H. (2017). The First Chloroplast Genome Sequence of *Boswellia sacra*, a Resin-Producing Plant in Oman. *PLoS One, 12*, e0169794.

King, R., & Zalucki, J. M. (2012). Potential inbreeding in a small population of a mass flowering species, Xanthorrhoea johnsonii (Xanthorrhoeaceae): is your mother my father? *American Journal of Plant Sciences, 3*, 303–312.

Kjølner, S., Såstad, S., Taberlet, P., & Brochmann, C. (2004). Amplified fragment length polymorphism versus random amplified polymorphic DNA markers: clonal diversity in *Saxifraga cernua*. *Molecular Ecology, 13*, 81–86.

Koch, P., Platzer, M., & Downie, B. R. (2014). RepARK—de novo creation of repeat libraries from whole-genome NGS reads. *Nucleic Acids Research, 42*, e80–e80.

Kramer, A. T., Ison, J. L., Ashley, M. V., & Howe, H. F. (2008). The paradox of forest fragmentation genetics. *Conservation Biology, 22*, 878–885.

Lacombe, T., & Thomas, M. R. (2006). Historical origins and genetic diversity of wine grapes. *Trends in Genetics, 22*(9), 511–519.

Laikre, L., Schwartz, M. K., Waples, R. S., & Ryman, N. (2010). Compromising genetic diversity in the wild: unmonitored large-scale release of plants and animals. *Trends in Ecology & Evolution, 25*, 520–529. https://doi.org/10.1016/j.tree.2010.06.013

Lamy, T., Laroche, F., David, P., Massol, F., & Jarne, P. (2017). The contribution of species–genetic diversity correlations to the understanding of community assembly rules. *Oikos, 126*, 759–771.

Lanfear, R., Calcott, B., Kainer, D., Mayer, C., & Stamatakis, A. (2014). Selecting optimal partitioning schemes for phylogenomic datasets. *BMC Evolutionary Biology, 14*, 82. https://doi.org/10.1186/1471-2148-14-82

Lawton-Rauh, A. (2008). Demographic processes shaping genetic variation. *Current Opinion in Plant Biology, 11*, 103–109.

Ledig, F. T., & Fryer, J. H. (1972). A pocket of variability in *Pinus rigida*. *Evolution, 26*, 259–266.

Lemenih, M., Feleke, S., & Tadesse, W. (2007). Constraints to smallholders production of frankincense in Metema district, North-western Ethiopia. *Journal of Arid Environments, 71*, 393–403.

Maki, M. (1999). RAPD markers revealed less genetic variation in the endangered plant Cerastium fisherianum var. molle than in its widespread congener C. fisherianum var. fisherianum (Caryophyllaceae). *Molecular Ecology, 8*, 145–150.

Moran, E. V., & Clark, J. S. (2011). Estimating seed and pollen movement in a monoecious plant: a hierarchical Bayesian approach integrating genetic and ecological data. *Molecular Ecology, 20*, 1248–1262.

Moran, E. V., & Clark, J. S. (2012). Between-site differences in the scale of dispersal and gene flow in red oak. *PLoS One, 7*, e36492.

Moran, P. (2002). Current conservation genetics: building an ecological approach to the synthesis of molecular and quantitative genetic methods. *Ecology of Freshwater Fish, 11*, 30–55.

Nisar, M., Ghafoor, A., Khan, M. R., Ahmad, H., Qureshi, A. S., & Ali, H. (2007). Genetic diversity and geographic relationship among local and exotic chickpea germplasm. *Pakistan Journal of Botany, 39*, 1575–1581.

Novaes, R., Rodrigues, J., & Lovato, M. (2009). An efficient protocol for tissue sampling and DNA isolation from the stem bark of Leguminosae trees. *Genetics and Molecular Research, 8*, 86–96.

Nybom, H. (2004). Comparison of different nuclear DNA markers for estimating intraspecific genetic diversity in plants. *Molecular Ecology, 13*, 1143–1155.

Ouborg, N. J., Angeloni, F., & Vergeer, P. (2010). An essay on the necessity and feasibility of conservation genomics. *Conservation Genetics, 11*, 643–653.

Parmentier, I., Duminil, J., Kuzmina, M., Philippe, M., Thomas, D. W., Kenfack, D., … Hardy, O. J. (2013). How Effective Are DNA Barcodes in the Identification of African Rainforest Trees? *PLoS One, 8*, e54921. https://doi.org/10.1371/journal.pone.0054921

Peakall, R., Ruibal, M., & Lindenmayer, D. B. (2003). Spatial autocorrelation analysis offers new insights into gene flow in the Australian bush rat, *Rattus fuscipes*. *Evolution, 57*, 1182–1195.

PÉRez-Collazos, E., & CatalÁN, P. (2007). Genetic diversity analysis and conservation implications for the Iberian threatened populations of the irano-turanian relict *Krascheninnikovia ceratoides* (Chenopodiaceae). *Biological Journal of the Linnean Society, 92*, 419–429. https://doi.org/10.1111/j.1095-8312.2007.00882.x

Pinares, A., González-Astorga, J., Vovides, A. P., Lazcano, J., & Vendrame, W. A. (2009). Genetic diversity of the endangered endemic *Microcycas calocoma* (Miq.) A. DC (Zamiaceae, Cycadales): Implications for conservation. *Biochemical Systematics and Ecology, 37*, 385–394. https://doi.org/10.1016/j.bse.2009.07.006

Ponce-Reyes, R., Nicholson, E., Baxter, P. W. J., Fuller, R. A., & Possingham, H. (2013). Extinction risk in cloud forest fragments under climate change and habitat loss. *Diversity and Distributions, 19*, 518–529. https://doi.org/10.1111/ddi.12064

Powell, W., Machray, G. C., & Provan, J. (1996). Polymorphism revealed by simple sequence repeats. *Trends in Plant Science, 1*, 215–222.

Pradeep Reddy, M., Sarla, N., & Siddiq, E. A. (2002). Inter simple sequence repeat (ISSR) polymorphism and its application in plant breeding. *Euphytica, 128*, 9–17. https://doi.org/10.1023/a:1020691618797

Quazi, M., Al-Riyami, W., Al-Jadeedi, S., Ambu-Saidi, H., Ali, A., Al-busaidi, H., … Hakkim, F. (2014). Conservation of omani frankincense (*boswellia sacra*, burseraceae) through plant tissue culture approach. *Journal of Science, 4*, 632–636.

Raffaelli, M., Mosti, S., & Tardelli, M. (2003). The Frankincense Tree (*Boswellia sacra* Flueck., Burseraceae) in Dhofar, southern Oman: field-investigations on the natural populations. *Webbia, 58*, 133–149. https://doi.org/10.1080/00837792.2003.10670749

Raffaelli, M., Mosti, S., & Tardelli, M. (2006). *Boswellia sacra* Flueck. (Burseraceae) in the Hasik area (Eastern Dhofar, Oman) and a list of the surrounding flora. *Webbia, 61*, 245–251. https://doi.org/10.1080/00837792.2006.10670804

Raffaelli, M., Tardelli, M., & Mosti, S. (2008). Preserving and restoring the frankincense tree (*Boswellia sacra*) at Wadi Doka: a work in progress. *A port in Arabia between Rome and the Indian Ocean (3rd C BC–5th C. AD). Khor Rori Report, 2*, 715–723.

Raju, A., Lakshmi, P. V., Ramana, K. V., & Chandra, P. H. (2012). Entomophily, ornithophily and anemochory in the self-incompatible Boswellia ovalifoliolata Bal. & Henry (Burseraceae), an endemic and endangered medicinally important tree species. *Journal of Threatened Taxa, 4*, 2673–2684.

Rakoczy-Trojanowska, M., & Bolibok, H. (2004). Characteristics and a comparison of three classes of microsatellite-based markers and their application in plants. *Cellular & Molecular Biology Letters, 9*, 221–238.

Ratnam, W., Rajora, O. P., Finkeldey, R., Aravanopoulos, F., Bouvet, J.-M., Vaillancourt, R. E., … Vinson, C. (2014). Genetic effects of forest management practices: global synthesis and perspectives. *Forest Ecology and Management, 333*, 52–65.

Rijkers, T., Ogbazghi, W., Wessel, M., & Bongers, F. (2006). The effect of tapping for frankincense on sexual reproduction in *Boswellia papyrifera*. *Journal of Applied Ecology, 43*, 1188–1195. https://doi.org/10.1111/j.1365-2664.2006.01215.x

Schaal, B. A., Hayworth, D. A., Olsen, K. M., Rauscher, J. T., & Smith, W. A. (1998). Phylogeographic studies in plants: problems and prospects. *Molecular Ecology, 7*, 465–474. https://doi.org/10.1046/j.1365-294x.1998.00318.x

Segelbacher, G., Cushman, S. A., Epperson, B. K., Fortin, M.-J., Francois, O., Hardy, O. J., … Manel, S. (2010). Applications of landscape genetics in conservation biology: concepts and challenges. *Conservation Genetics, 11*, 375–385.

Slatkin, M. (1987). Gene flow and the geographic structure of natural populations. *Science, 236*, 787–792.

Smouse, P. E., Peakall, R., & Gonzales, E. (2008). A heterogeneity test for fine-scale genetic structure. *Molecular Ecology, 17*, 3389–3400.

Smulders, M., Cottrell, J., Lefèvre, F., Van der Schoot, J., Arens, P., Vosman, B., … Castiglione, S. (2008). Structure of the genetic diversity in black poplar (*Populus nigra* L.) populations across European river systems: consequences for conservation and restoration. *Forest Ecology and Management, 255*, 1388–1399.

Sreekumar, V., & Renuka, C. (2006). Assessment of genetic diversity in Calamus thwaitesii BECC. (Arecaceae) using RAPD markers. *Biochemical Systematics and Ecology, 34*, 397–405.

Sunnichan, V., Mohan Ram, H., & Shivanna, K. (2005). Reproductive biology of *Boswellia serrata*, the source of salai guggul, an important gum-resin. *Botanical Journal of the Linnean Society, 147*, 73–82.

Teshome, M. (2013). *Structure and composition of woody plants in Boswellia dominated woodland of Western Ethiopia*. MSc Thesis. Wageningen University: The Netherlands.

Teshome, M., Eshete, A., & Bongers, F. (2017). Uniquely regenerating frankincense tree populations in western Ethiopia. *Forest Ecology and Management, 389*, 127–135. https://doi.org/10.1016/j.foreco.2016.12.033

Tolera, M., Sass-Klaassen, U., Eshete, A., Bongers, F., & Sterck, F. J. (2013). Frankincense tree recruitment failed over the past half century. *Forest Ecology and Management, 304*, 65–72.

Vekemans, X., & Hardy, O. J. (2004). New insights from fine-scale spatial genetic structure analyses in plant populations. *Molecular Ecology, 13*, 921–935.

Verde, I., Abbott, A. G., Scalabrin, S., Jung, S., Shu, S., Marroni, F., … Cattonaro, F. (2013). The high-quality draft genome of peach (*Prunus persica*) identifies unique patterns of genetic diversity, domestication and genome evolution. *Nature Genetics, 45*, 487.

Chapter 7
Boswellia sacra Plastid Genome Sequencing and Comparative Analysis

Chloroplast DNA

Chloroplasts, an active metabolic centre of the cell, and its circular, independent genome are composed of nearly 130 genes of approximately 72–217 kb in length (Corriveau & Coleman, 1988; Zhang & Liu, 2003). The conserved nature of the chloroplast DNA (cpDNA) in terms of their gene content, double-stranded DNA molecule and gene order makes it the best choice for studying evolution in the plant kingdom (Moore, Bell, Soltis, & Soltis, 2007). With the development of sequencing technologies, especially the next-generation sequencing technologies, there has been an abrupt increase in chloroplast genome sequencing, which provides useful information for understanding the functional and structural evolutionary prospects of non-model ecologically and economically important plants (Jansen et al., 2006; Moore et al., 2007; Moore, Soltis, Bell, Burleigh, & Soltis, 2010; Wang, Wuyun, Du, Wang, & Cao, 2016). Recently, a number of studies have explored the complete genome sequence applications of economically and ecologically important tree species, such as *Eucommia ulmoides* (Shang, Du, Yin, & Zhang, 2016), *Populus cathayana* (Du et al., 2015), *Quercus spinosa* (Williams, Boykin, Howell, Nevill, & Small, 2015), *Acacia ligulata* (Li et al., 2016), *Pinus armandii* (Huang, Matzke, & Matzke, 2013), *Cocos nucifera* (Su, Hogenhout, Al-Sadi, & Kuo, 2014), *Citrus aurantifolia* (Martin, Baurens, Cardi, Aury, & D'Hont, 2013), *Musa acuminata* (Nystedt et al., 2013), Norway spruce (Raffaelli, Mosti, & Tardelli, 2003), etc. In recent years, chloroplast genome sequencing has been elucidated for numerous important trees to obtain a direct understanding of their gene structure, which ultimately facilitates targeted conservation and different propagation strategies (Jansen et al., 2006; Moore et al., 2007).

A. Al-Harrasi et al., *Biology of Genus Boswellia*,
https://doi.org/10.1007/978-3-030-16725-7_7

The chloroplast present in the plant cell was proposed to originate through the endosymbiosis of cyanobacteria and plant cells (Jin & Daniell, 2015; Qian et al., 2013; Sugiura, 1992). The basic key function of the chloroplast includes the generation of energy for plant growth and development through photosynthesis (Ohyama et al., 1986; Xie, Michaeli, Peled-Zehavi, & Galili, 2015). Despite this function, chloroplasts also assist cells in fatty acid, starch, pigment and amino acid biosynthesis processes (Jensen, 2013). Currently, approximately 800 complete chloroplast (cp) genomes have been submitted to NCBI, which is publicly available (http://www.ncbi.nlm.nih.gov/genome). The majority of the sequenced chloroplast genomes belong to the culturally and economically important plants (Cosner, Raubeson, & Jansen, 2004). The chloroplast genome has the ability to insert a large number of foreign genes, and it can also be easily and safely engineered compared to its nuclear genome. The genomic analysis of this organelle can provide useful and important knowledge used in the development of plant sources, which could be efficient at counteracting the environmental stresses and further improve the human interests related to the target plant (Jansen & Ruhlman, 2012). Chloroplast genome sequencing provides us with more information that can also help us to understand the major biosynthesis pathways and function. Due to the maternal inheritance of the chloroplast genome, it is widely used in plant systematics (Saski et al., 2007). Due to the highly conserved structure of this genome, it is widely used in designing nucleotide markers for plant barcoding and population-wide genetic studies.

Molecular and General Features of the *Boswellia sacra* Chloroplast Genome

Chloroplast genomes of *B. sacra* were sequenced on Ion Torrent PGM, and a total of 286.8 Mb in data were generated after assembling the sequences. The reads were de novo assembled (373,248 reads, 112,936,195 bp) to show almost 730X of the chloroplast genome coverage. To obtain a high-quality cp genome from *B. sacra*, another round of chloroplast DNA was extracted, and libraries were prepared and sequenced on Ion Torrent PGM. The resulting data were sufficient to generate a long single-molecule sequence representing the complete *Boswellia sacra* cp genome. Some previously reported chloroplast genomes from the order Sapindales such as *Citrus sinensis* and *Azadirachta indica* were used as a reference for the newly generated de novo assembled chloroplast genome sequences of *B. sacra*. The gap in the de novo assembly sequences was filled from the available raw reads and the cp genome of *B. sacra*, and then the de novo assemblies were remapped with the previously reported reference genome retrieved from NCBI. Error-free sequences of *B. sacra* were submitted to GenBank under the accession number KT934315.

The complete cp genome of *Boswellia sacra* is a closed, single, circular molecule of 160,543 bp in size consisting of a pair of inverted repeats (IRa and IRb) of 26,763 bp separated by LSC (88,055 bp) and SSC regions (18,962 bp) (Fig. 7.1). Based on the annotation, 114 different genes were identified, with 83 protein-coding

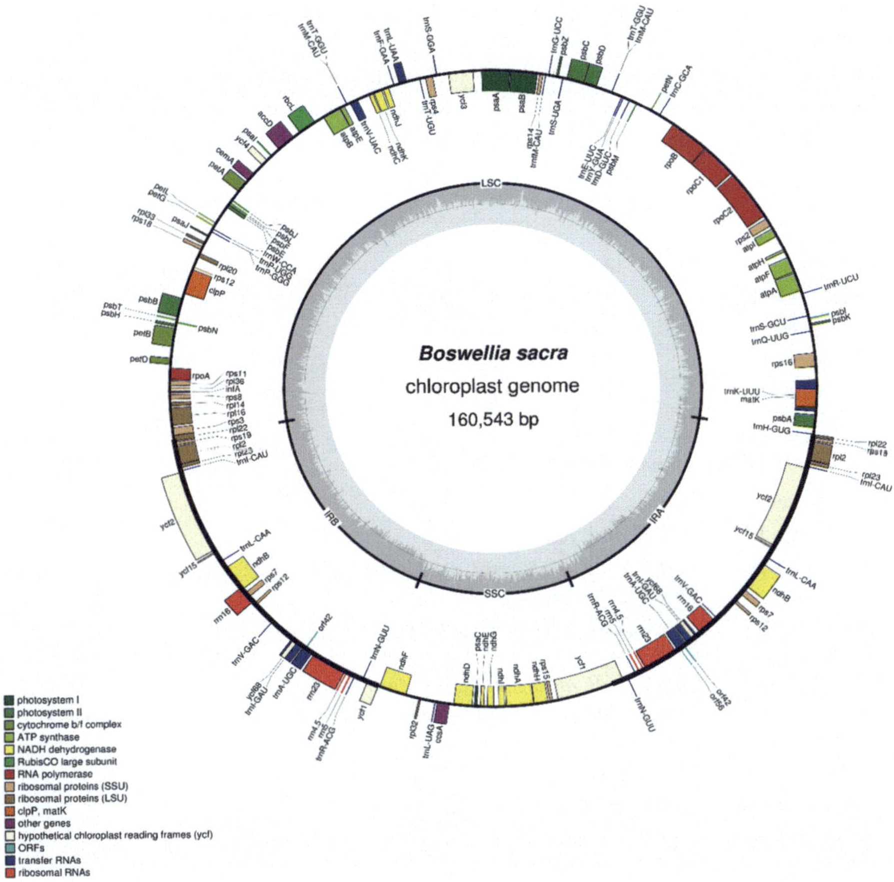

Fig. 7.1 Gene map of the *Boswellia sacra* chloroplast genome. A pair of thick lines in the inside circle represents the inverted repeats (IRa and IRb; 26,763 bp each), and they separate the large single copy region (LSC; 88,055 bp) from the small single copy region (SSC; 18,962 bp). The genes drawn inside the circle are transcribed clockwise, while those drawn outside the circle are transcribed counterclockwise

genes, 27 tRNA and 2 rRNA genes, respectively (Table 7.1). The duplicated gene observed in the IR region is 24. The complete *B. sacra* cp genome shared high similarity with the previously reported *Azadirachta indica* and *Citrus sinensis* genomes (160,129 bp and 160,737, respectively). However, the complete cp genome size of *B. sacra* is larger than that of *Citrus aurantifolia* (159,893 bp), which belongs to the same Sapindales order. The *B. sacra* cp genome size was found to be higher than that of other arid land plants, e.g. the *Acacia ligulata* (158,724 bp) cp genomes were reported in previous studies (Hudson & Gardner, 1988; Li et al., 2016; Martin et al., 2013). Although the total lengths differed, the structure was consistent with previously reported restriction mapping (Jansen, Wojciechowski, Sanniyasi, Lee, & Daniell, 2008).

Table 7.1 Summary of the chloroplast genome characteristics in *Boswellia sacra* with related cp genomes

Attribute	*B. sacra*	*A. indica*	*C. wightii*	*C. limon*	*C. maxima*	*C. sinensis*
Size (bp)	160,543	160,737	156,064	160,101	160,133	160,129
GC content (%)	37.6	37.5	38	38.5	38.5	38.5
LSC (bp)	88,055	88,136	–	87,720	88,043	87,744
SSC (bp)	18,962	18,635	–	18,393	18,400	18,393
IR (bp)	26,763	26,983	–	26,994	26,845	26,996
Total genes	142	131	123	138	134	140
Protein-coding genes	93	86	86	93	89	87
rRNA genes	8	8	8	8	8	8
tRNA genes	39	37	29	37	37	45
Genes with introns	**16**	**17**	**12**	**20**		**17**

Table 7.2 *Boswellia sacra* gene list

Group of gene	Genes
Photosystem I	*psaC, psaI, psaJ, psaA, psaB*
Photosystem II	*psbE, psbF, psbH, psbI, psbJ, psbA, psbB, psbC, psbD, psbK, psbL, psbM, psbN, psbT, psbZ*
Cytochrome b/f complex	*petA, petB, petD, petG, petL, petN*
ATP synthase	*atpA, atpB, atpE, atpF, atpH, atpI*
NADH dehydrogenase	*ndhA, ndhB, ndhC, ndhD, ndhE, ndhF, ndhG, ndhH, ndhI, ndhJ, ndhK*
RubisCO	*Large subunit rbcL*
RNA polymerase	*rpoA, rpoB, rpoC1, rpoC2*
Ribosomal proteins (SSU)	*rps2, rps3, rps4, rps7, rps8, rps11, rps12, rps12_3end, rps14, rps15, rps16, rps18, rps19*
Ribosomal proteins (LSU)	*rpl2, rpl14, rpl16, rpl20, rpl22, rpl23, rpl32, rpl33, rpl36*
Other genes	*clpP, matK, accD, ccsA, infA, cemA*
Hypothetical chloroplast reading frames	*ycf1, ycf2, ycf3, ycf4, ycf15, ycf68*
ORFs	*orf42, orf56*
Transfer RNAs	*trnA-UGC, trnC-GCA, trnD-GUC, trnE-UUC, trnF-GAA, trnG-UCC, trnH-GUG, trnI-CAU, trnI-GAU, trnK-UUU, trnL-CAA, trnL-UAG, trnM-CAU, trnN-GUU, trnP-GGG, trnP-UGG, trnQ-UUG, trnR-ACG, trnR-UCU, trnS-GCU, trnS-GGA, trnT-GGU, trnV-GAC, trnV-UAC, trnW-CCA, trnY-GUA*
Ribosomal RNAs	*rrn4.5, rrn5, rrn16, rrn23*

All the possible genes involved in the photosynthesis process were identified (Tables 7.1 and 7.2). There are 20 genes responsible for photosystems I and II and another 17 genes that play an important role in ATP synthase and NADH dehydrogenase followed by 6 genes with a function in the cytochrome b/f complex,

Table 7.3 Chloroplast genes found in *B. sacra* and their functions and products

Gene (s)	Role	Product
acc	Fatty acid metabolism	C-type cytochrome synthesis protein; acetyl-CoA carboxylase biotin carboxyl carrier protein; acetyl-CoA carboxylase beta subunit; acyl carrier protein
ycf	Conserved hypo	Hypothetical protein complex; ABC transporter; chloroplast
atp	Energy metabolism	Adenosine triphosphate synthase
ccs	Biosyn of cofactors	Cytochrome c biogenesis protein
cem	Transport membrane	Chloroplast envelope membrane protein
clp	Cellular processes	ATP-dependent clp protease ATP-binding subunit
inf	Hypothetical/ uncharacterized	Translation initiation factors
lsc	Long single copy	Adenosine triphosphate synthase
Mat	Hypothetical/ uncharacterized	Maturase K
ndh	Energy metabolism	NADH-plastoquinone reductase
pet	Energy metabolism	Cytochrome b6 complex protein family
psa	Photosynthesis	Photosystem I reaction centre complex
psb	Photosynthesis	Photosystem II reaction complex
RbcL	Photosynthesis	Ribulose bisphosphate carboxylase large chain
rpl	Translation	Ribosomal protein involved in translation
rpo	Transcription	DNA-directed RNA polymerase alpha chain
rps	Translation	Ribosomal protein
rrn	rRNA	Plastid rRNA
trn	tRNA	Plastid tRNA

2 ORFs, *rbcL* and some other important genes related to the ATP-dependent protease/ATP-binding subunit (*clpP*), envelope membrane protein (*cemA*), subunit of acetyl-Co-A carboxylase (*accD*), translational initiation factor (*infA*), maturase K (*matK*) and c-type cytochrome synthesis gene (*ccsA*) (Tables 7.2 and 7.3). In the *B. sacra* cp genome, the presence of a chloroplast-encoded *clpP* indicated that this gene is necessary for cell survival. However, this assumption requires scientific verifications, but the presence of *clpP* shows that there is a better system to combat high heat and drought stress in *B. sacra*, while other species, e.g. *Cicer arietinum*, *Pisum sativum* and *Lathyrus sativus*, have lost an intron from the *clpP1* and *rps12* genes (Sveinsson & Cronk, 2014). The *clpP1* gene is identified in other species, namely, *Trifolium subterraneum* (Yang, Yang, Li, Yang, & Li, 2013), *Acacia ligulata* (Li et al., 2016), Actinidiaceae (Yap et al., 2015), *Wollemia nobilis* (Shaw, Lickey, Schilling, & Small, 2007) and *Camellia* sp. (Yao et al., 2015).

Table 7.4 Differences between the *B. sacra* and *A. indica* cp genomes

Indel				
		Length (bp)	Count	
		1	2743	
		2–10	1216	
		11–21	12	
	Sum	6353	3971	Percentage[a]: 3.96%
Substitution				
		Type	Count	
		A<->T	1571	
		C<->G	837	
		A<->C	1416	
		T<->C	1828	
		A<->G	1777	
		T<->G	1500	
	Sum		8929	Percentage[a]: 5.56%
10 most divergent intergenic regions				
		Region	Length[b] (bp)	Pairwise distance
		trnH-GUG-psbA	457	0.47
		accD-psaI	746	0.32
		atpH-atpI	1167	0.31
		psbZ-trnG-UCC	546	0.30
		trnK-UUU-rps16	1034	0.25
		petN-psbM	1139	0.25
		ycf4-cemA	942	0.25
		ccsA-ndhD	307	0.24
		trnR-UCU-atpA	198	0.23
		trnL-UAG-ccsA	111	0.21

[a]Relative to the length of *B. sacra*
[b]Length in *B. sacra*

Comparisons of the *Boswellia sacra* cp Genome with Related Species

B. sacra cp genomes were compared with those of *Commiphora wightii*, which belongs to the same family, Burseraceae, and to those of *A. indica*, *Citrus limon*, *C. maxima* and *C. sinensis*, from the same Lamiales order. These genomes were generally compared for their general characteristics (Table 7.4). Based on the current comparison, high similarities were noted in these three genomes. The size of the *B. sacra* cp genome (160,543 kb) was smaller than that of *A. indica* (160,737 kb) and greater than that of *C. wightii* (156,064 kb), *Citrus limon* (160,101 kb), *C. maxima* (160,133 kb) or *C. sinensis* (160129). The LSC consists of 88,055 bp in *B. sacra*;

88,136 bp in *A. indica*; 87,720 in *C. limon*; 88,043 bp in *C. maxima*; and 87,744 bp in *C. sinensis*.

Similarly, the SSC region found in the *B. sacra* and comparison genomes were 18,962 bp in *B. sacra*, 18,635 bp in *A. indica*, 18,393 bp in *C. limon*, 18,400 bp in *C. maxima* and 183,939 bp in *C. sinensis*. The IR regions found in these genomes were 26,793 bp in *B. sacra*, 26,983 bp in *A. indica*, 26,994 bp in *C. limon*, 26,845 bp in *C. maxima* and 26,996 bp in *C. sinensis*. The size of the *B. sacra* genome is similar to the size of the typical angiosperm genome, which is less than the *Eucommia ulmoides* chloroplast genome at 163,341 bp (Wang, Wuyun, et al., 2016) and greater than that of *Salvia miltiorrhiza* at 151,328 bp (Qian et al., 2013). The GC content present in the cp genome of *B. sacra* (37.61%) was generally lower than the GC contents in that of *Commiphora wightii* (38%), *Citrus limon* (38.5), *C. maxima* (38.5) and *C. sinensis* (38.5) cp genome and higher than the *A. indica* cp genome (37.5%). The total genes present in *B. sacra* (142) were higher than the compared genomes, at *A. indica* (131), *C. wightii* (123), *Citrus limon* (138), *C. maxima* (134) and *C. sinensis* (140). Similarly, there were more protein-coding genes present in *B. sacra* (92) than in the compared genomes from *A. indica* (86), *C. wightii* (86), *Citrus limon* (93), *C. maxima* (89) and *C. sinensis* (87). The number of genes is almost similar to that of *Haberlea rhodopensis*, which contains 137 genes (Ivanova et al., 2017). The numbers of rRNA genes (8) were found to be the same in *B. sacra* and all the other genomes. The tRNA genes present in the *B. sacra* numbered (39), while the highest number was in the *C. sinensis* (45) chloroplast genome. There were 16 intron-containing genes found in *B. sacra*, 17 in *A. indica*, 12 in *C. wightii*, 20 in *C. limon* and 17 in *C. sinensis*. Intergenic regions (33.5%), introns (9.6%) and genic regions (57.5%) were found in the cp genome of *B. sacra* (Tables 7.1 and 7.2). Upon pairwise alignment, the *B. sacra* chloroplast genome showed a sequence divergence of 9.52% (Tables 7.3, 7.4, and 7.5), and moreover, the *B. sacra* cp genome contains a total of 3971 *indels* (3.96%) and 8929 substitutions (5.56%). Variations in the sizes of the LSCs and SSCs from the cp genomes were also observed in *B. sacra* and *A. indica*, which reflects the larger index in each region of the genome. Although there were many divergent regions present during the comparison of these genomes, some regions possess a more obvious divergence, and here, we will consider the ten highly divergent intergenic regions to be *rnH-GUG-psbA* (457 bp, 40.54% divergence), the spacer between these, *accD-psaI* (746 bp, 37.27% divergence), *atpH-atpI* (1167 bp, 33.94% divergence), *psbZ-trnG-UCC* (546 bp, 28.14% divergence), *trnK-UUU-rps16* (1034 bp, 26.40% divergence), *petN-psbM* (1139 bp, 36.21% divergence), *ycf4-cemA* (942 bp, 26.69% divergence), *ccsA-ndhD* (307 bp, 25.24% divergence), *trnR-UCU-atpA* (198 bp, 28.97% divergence) and *trnL-UAG-ccsA* (111 bp, 33.05% divergence). These findings indicate that a deletion might have occurred in this region of *A. indica* after the divergence of *A. indica* and *B. sacra*.

The *B. sacra* cp genomes were highly conserved regarding the *indel* site; however, only 25 total variable regions were identified. The *ycf1* region showed high variability, with its 4354 *indel* sites with an average *indel* length of 174 bp, and the *atpA* gene also exhibited variability and contained 124 indel sites. However, the

Table 7.5 List of simple sequence repeats

Unit	Size (bp)	Total Number of SSRs	Starting position[a]
AT	10	5	**21,733** (*rpoC2*); 50,519; 63,260; 120,682; **122,312** (*ndhD*)
	12	2	72,021; 118,991
TA	10	1	**50,086**
	12	1	49,380
TC	10	1	**65,118** (*cemA*)
AAG	12	1	**98,114**
ATA	12	3	**52,097**; 57,925 (*atpB*); 70,908
CTT	12	1	**150,474**
TAT	12	1	50,027
TTA	12	2	**54,843**; 131,240 (*ycf1*)
AAAT	12	1	6383
AATT	12	1	117,168
ATAG	12	1	34,089
CATT	12	1	**131,003** (*ycf1*)
CTTT	12	1	51,864
TATC	12	1	38,353
TTTC	12	1	125,090
ATATGA	18	1	118,904

[a]The SSR-containing coding regions are indicated in parentheses. SSRs that are identical in the *A. indica* chloroplast genome are highlighted in bold

highest *indel* diversity per site was recorded for *rpl2* (Table 7.6). The *B. sacra* and *A. indica* cp genomes were compared, and the average pairwise sequence divergence levels were calculated among 76 common genes (Table 7.7). Among these, the *B. sacra* cp genome showed a 0.066 average sequence divergence. Furthermore, *rpl32 atpA*, *rps12*, *rpl2*, *ycf1 rpl20*, *infA*, *ndhF*, *rpl22* and *rps8* showed high sequence divergence. The *atpA* gene (0.852) sequence divergence was the highest among them all, followed by *rps11* (0.301).

The CBOL group most recently suggested *trnH/psbA* as a putative marker for the identification of land plants. This region has been tested in many plant species such as *Araucaria* (Shaw et al., 2007; Yao et al., 2015; Yap et al., 2015). The *trnH/psbA* is a non-coding region of the cp genome that is highly variable in sequence length and is flanked by coding regions from both sides, i.e. *psbA*, which encodes photosystem II protein D1 and *trnH-GUG*, which functions in histidine translation (Jansen et al., 2006; Moore et al., 2007). The *B. sacra* region was recorded to be 457 bp in length. Previously, Yap et al. (2015) suggested that the *trnH/psbA* region was missing in *Agathis dammara* compared with *Wollemia nobilis*. However, the

Table 7.6 List of long repeat sequences

Repeat size	Type[a]	Start position of 1st repeat	Start position the repeat found in another region	Location[b]	Region
30	P	9578	48,299	***trnS-GCU, trnS-GGA***	LSC
30	D	30,965	31,304	IGS (*petN-psbM*)	LSC
30	D, P	95,755	95,773, 152,797, 152,815	***ycf2***	IR
31	D	50,042	50,183	IGS (*trnT-UGU-trnL-UAA*)	LSC
32	D	15,474	15,696, 16,029	IGS (*atpH-atpI*)	LSC
34	P	5863	5863	Intron (*rps16*)	LSC
36	D, P	103,143	125,725, 145,421	**IGS (*rps12-trnV-GAC*), intron** (*ndhA*), **IGS** (*trnV-GAC-rps12*)	IR, SSC
40	D	64,428	64,797	IGS (*ycf4-cemA*)	LSC
42	D	49,991	50,135	IGS (*trnT-UGU-trnL-UAA*)	LSC
42	D	64,386	64,759	IGS (*ycf4-cemA*)	LSC
52	P	444	444	IGS (*trnH-GUG-psbA*)	LSC
56	P	31,815	31,815	**IGS (*petN-psbM*)**	LSC
65	D	15,507	16061[c]	IGS (*atpH-atpI*)	LSC
69	D	31,107	31,522	IGS (*petN-psbM*)	LSC
76	D	41,575	43,799	*psaB, psaA*	LSC
109	D	64,274	64,648	IGS (*ycf4-cemA*)	LSC
155	D	15,417	15,639	IGS (*atpH-atpI*)	LSC
204	D	15,729	16061[c]	IGS (*atpH-atpI*)	LSC
251	D	62,442	62,769	*IGS* (*accD-psaI*), *psaI*	LSC

[a]*D* direct repeat, *P* palindrome-inverted repeat
[b]*IGS* intergenic spacer region. Sequences conserved in the *A. indica* chloroplast genome are highlighted in bold
[c]The 65 bp sequence is a truncated copy of a 204 bp sequence

presence of this *indel* was reported in 19 *Araucaria* species including *W. nobilis* based on a detailed BLAST analyses (Yap et al., 2015). In addition, *psbA* has a function against oxidative stress during drought in photosystem II. Huo et al. (Leliaert & Lopez-Bautista, 2015) concluded that *psbA* gene overexpression during oxidative stress increased tobacco plant tolerance to drought stress. The presence of the *psbA* gene in the *B. sacra* cp genome suggests a strong ability to counteract oxidative stress, because *B. sacra* frequently faced low water stress; however, this question requires more investigation.

Table 7.7 Distribution of tandem repeats in the *B. sacra* chloroplast genome

Indices	Repeat length	Copy number	Percent indels	Percent matches	Location
347–387	16	2.6	11	88	*trnH/psbA* (IGS)
4566–4594	12	2.4	0	100	*trnK/rps16* (IGS)
4819–4874	19	2.9	5	78	*trnK/rps16* (IGS)
7784–7814	16	1.9	0	100	*rps16/trnQ*(IGS)
8690–8726	18	2.1	0	94	*trnQ/psbK*(IGS)
9763–9828	27	2.4	5	84	*trnS/trnR* (IGS)
9783–9831	14	3.6	5	83	*trnS/trnR* (IGS)
9779–9867	27	3.5	13	85	*trnS/trnR* (IGS)
9863–9892	14	2.1	0	100	*trnS/trnR* (IGS)
11,501–11,572	31	2.5	18	81	*trnR/atpA*(IGS)
11,504–11,573	25	2.7	17	76	*trnR/atpA*(IGS)
11,535–11,583	22	2.4	12	80	*trnR/atpA*(IGS)
16,355–16,385	14	2.2	0	94	*atpH/atpI* (IGS)
29,582–29,606	12	2.1	0	100	*rpoB/trnC*(IGS)
31,751–31,776	13	2	0	100	*petN/psbM*(IGS)
49,997–50,046	18	2.9	11	82	*trnT/trnL*(IGS)
60,254–60,278	12	2.1	0	100	*rbcL/accD*(IGS)
62,281–62,429	76	2	5	88	*accD*(CDS)
67,916–67,954	19	2.1	4	90	*psbJ/psbL*(IGS)
69,384–69,454	12	6.1	18	72	*psbE/petL*(IGS)
72,176–72,214	21	1.9	5	94	*rps18* (CDS)
73,287–73,312	13	2	0	100	*rpl20/rps12*(IGS)
95,746–95,802	18	3.2	0	97	*ycf2*(CDS)
112,196–112,261	32	2.1	0	97	*rrn 4.5/rrn5*(IGS)
113,767–113,809	20	2.2	0	100	*trnN/ycf1*(IGS)
119,576–119,605	15	2	0	93	*ccsA*(CDS)
129,059–129,085	13	2.1	0	100	*rps15/ycf1*(IGS)
134,790–134,832	20	2.2	0	100	*ycf1*(CDS)
136,338–136,403	32	2.1	0	97	*rrn5/rrn4.5*(IGS)
152,797–152,853	18	3.2	0	97	*ycf2*(CDS)

Analysis of Repetitive Sequences

In the chloroplast genome, repetitive sequences play an important key role in the rearrangement and stabilization of the genome, and they provide important information for understanding the evolutionary history of plant species (Lemieux, Otis, & Turmel, 2014; Qin et al., 2015). To understand evolutionary history, mononucleotide length polymorphisms have been used, and they show high variations in the numbers in the chloroplast genomes (Yao et al., 2015). In the current study, we observed a total of 109 SSR loci in the cp genome, which made up 1256 bp (ca. 0.78%) of the overall sequence (Table 7.5). In the cp genome of *B. sacra*, the intergenic region containing most of the SSRs was also located in the coding areas, i.e. the ycf1 and the rpoC2 genes, which are widely used to determine the population diversity in many plant groups (Qin et al., 2015). In our present study, we have found 39 large repeats with lengths of more than 30 bp in the *Boswellia sacra* cp genome (Table 7.6).

Previously, tandem repeats have been identified in conifer cp genomes, but mechanisms involved in the formation of these repeats still remain unclear. Nevertheless, tandem repeats were found to be involved in cp DNA rearrangement (Addisalem, Esselink, Bongers, & Smulders, 2015; Yang et al., 2016). A total of 30 tandem repeats were found based on the result of a tandem repeat analysis (Table 7.7). The coding gene *accD* was determined to contain the largest repeats (76 bp), whereas the smallest was found in intergenic spacer region *trnK/rps16* (12 bp), with an average copy number of 2.4. The percent range of *indels* was from 0 to 18 (Table 7.4). A large number of these repeats were present in the intergenic spacers; although in the coding regions of *trnS-GCU*, *trnS-GGA*, *ycf2*, *psaB* and *psaA*, five repeats were found. In *A. indica*, eight long repeats were identified, which clearly suggest that these repeats are common to Sapindales. Addisalem et al. (Yang et al., 2016) developed a 46 SSR marker set for the first time and found that the genome size of *B. papyrifera* is 705 Mb. This was the first repeat representing genomic study for this genus. However, the existing knowledge about the genomics and transcriptomics of this keystone species remains limited.

The single-nucleotide polymorphism (SNP) maximum likelihood method of natural selection, codon-by-codon, was conducted to count the total number of inferred synonymous (s) and non-synonymous (n) substitutions present along with the numbers of sites estimated to be synonymous (s) and non-synonymous (n) (Muse & Gaut, 1994). These estimates are assumed by using the joint maximum likelihood reconstructions of ancestral states under a Muse-Gaut model (Seo & Kishino, 2008) of nucleotide substitution. In the genome of *B. sacra*, *A. indica* and *Citrus sinensis*, a total of 8301 positions were noted. For the codons, TCT, GTT, GGA and CGT showed high n, s, n and *dN-dS*. However, the substitutions among cp genomes were significant as indicated by the *P* values of less than 0.05. The normalized *dN/dS* ratio was calculated for the three cp genomes, which showed a positive

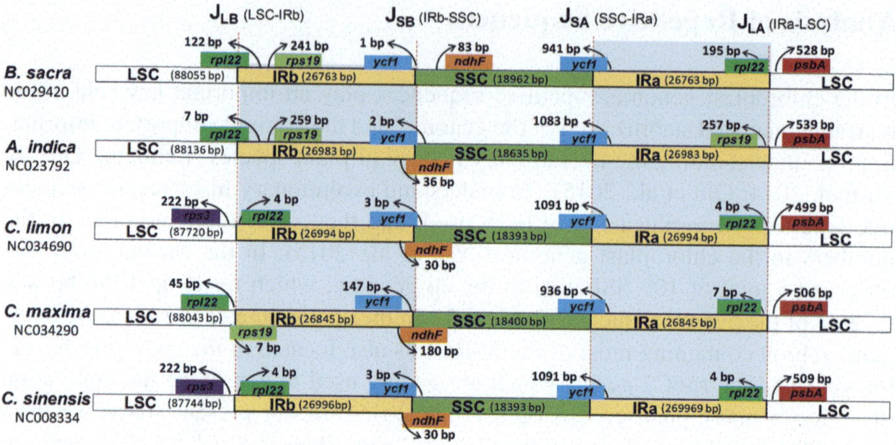

Fig. 7.2 Comparison of the border positions for the LSC, SSC and IR regions in the *B. sacra* chloroplast genome and a comparison with related species

selection ranging from 0 to 10.17. In the three compared cp genomes, 14,249 segregating sites were noted with a nucleotide diversity of 0.063. The nucleotide frequencies were 30.59% (A) followed by 31.21% (T/U), 19.49% (C) and 18.71% (G). The rate ration, transition and transversion were $k1 = 2.149$ (purines) and $k2 = 2.072$ (pyrimidines). The overall transition/transversion bias is ($R = 0.996$). The results showed that the species, namely, *B. sacra*, *A. indica* and *Citrus sinensis*, exhibited few changes in the structural organization of the chloroplast genomes over a long period. These results explain how these species evolved at very low rates. The same types of perspectives were illustrated for the cp genomes of five *Quercus* species (Muse & Gaut, 1994). The estimation of synonymous and non-synonymous substitution rates facilitate and assist in explaining the dynamics of molecular evolution (Ohtani, Yamamoto, & Akimitsu, 2002).

Junction Characteristics of the Chloroplast Genome

The contraction and expansion that occur frequently in the junction present in the chloroplast genome are primarily responsible for the size variation of the chloroplast genome (Wang et al., 2008). The four junctions present in the chloroplast genome of *B. sacra* were compared with the junctions in *A. indica*, *C. limon*, *C. maxima* and *C. sinensis* (Fig. 7.2). The overall compositions of these regions were conserved in the compared genomes. At the J_{LB} (LSC-IRb) junction, the *rpl22* gene is present in *B. sacra* and *A. indica*, and in *B. sacra*, it is present as 122 bp in the IRb region, while in *A. indica*, it is represented by 7 bp in the IRb region. In *C. limon*, the *rpl22* gene completely resides within the IRb region bp away from the start of

J_{LB} junction. In *C. maxima*, the *rpl22* gene is present 45 bp away from the J_{LB} junction in the LSC region, while in *C. sinensis*, the location of the *rpl22* gene is similar to that of the *C. limon* cp genome. Similarly, the *rpl22* gene is also present in the IRa region in all the compared genomes except in *A. indica*. At the J_{SA} junction, the *ycf1* gene is present in all the genomes, while at the J_{SB} junction, the ycf1 gene is present in *A. indica*, *C. limon* and *C. sinensis*, but it was absent at the J_{SA} junction in the *B. sacra* and *C. maxima* chloroplast genomes. The *ndhF* gene was found at the J_{SB} junction in all the compared genomes except that of *B. sacra*, in which it is completely contained within the SSC region 83 bp away from the J_{SB} junction. The *psbA* gene was completely contained within the LSC regions of all the compared genomes. In *B. sacra* and *A. indica*, the rps19 gene was present in the IRb region, while in *C. maxima*, it was found at the J_{LB} junction (Fig. 7.2). These contractions and expansions of the junctions show similar results to those of previously reported cp genomes from (Saina, Li, Gichira, & Liao, 2018). However, some of the deviations in sequences might be due to boundary contraction and expansion between the boundaries of the IR and single copy regions among different plant species as reported by Wang et al. (Wang, Wuyun, et al., 2016).

Analysis of Sequence Divergence

Using VISTA, the chloroplast genomes of *B. sacra* were plotted to compare the sequence divergence present in the chloroplast genome and then compared with those of *C. wightii*, *A. indica*, *C. limon*, *C. maxima* and *C. sinensis* (Fig. 7.3). Upon the analysis of the sequence divergence, it was revealed that there is more variation in the non-coding region of the chloroplast genome than in the coding region. The *B. sacra* cp genome was used as a reference for measuring the sequence divergence in the compared genomes. The matK gene in *B. sacra* and *C. wightii* shows no significant sequence divergence, while a small sequence divergence was recorded in the *A. indica*, *C. limon*, *C. maxima* and *C. sinensis* chloroplast genomes. Some clearly sequence-divergent regions were present in the chloroplast genome in comparison with the related genome as *rps16*, *atpF*, *rpoC1*, *rpoC2*, *ycf3*, *accD*, *clpP*, *petB*, *ycf1*, *ndhF*, *ccSA*, *ndhA*, *ndhB* and *ycf2*. Moreover, some variations were also observed in the RNA regions. Overall, the ycf 1 regions were found to be the most divergent regions present in these genomes. The sequence divergence in similar regions such as *matK*, *ycf3-psaA*, *clpP*, *accD*, *atpF*, *rpoC1*, *petA-psbJ*, *ycf1-rps15*, *rps19* and *ndhF* was reported previously for various cp genomes (Asaf et al., 2017; Qian et al., 2013).

Phylogenetic Analysis of *B. sacra*

Using small locus barcoding for plant phylogenetics and studying evolutionary relationships involve certain limitations compared to using the complete chloroplast genome (Li, Feng, Wang, Liu, & Lv, 2014; Percy et al., 2014). The use of the

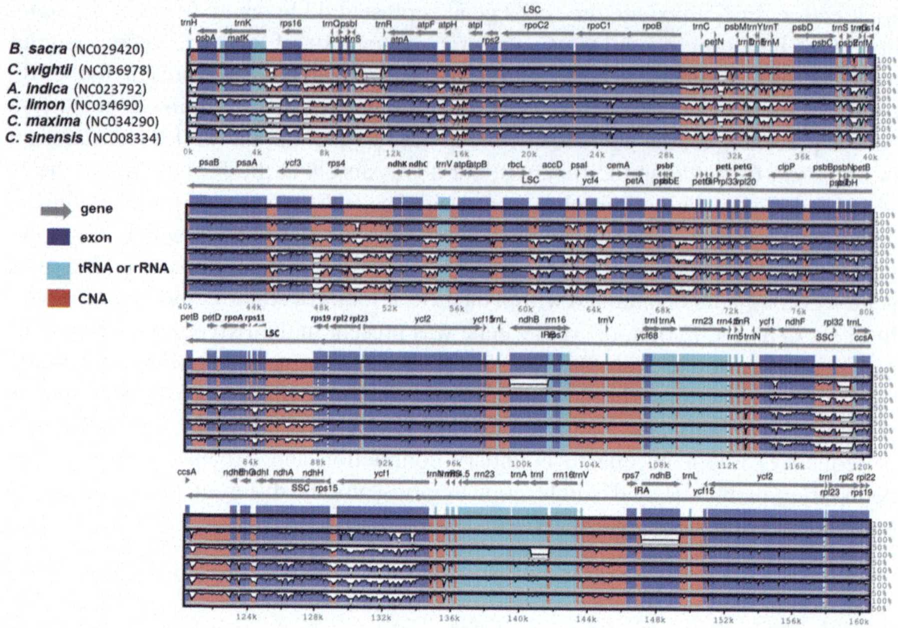

Fig. 7.3 Visualized alignment of chloroplast genome sequences from *B. sacra* cp genome with those of related species. VISTA-based similarity graphical information portraying the sequence identity of *B. sacra* with reference to the *A. indica* cp genomes. Thick black lines show the inverted repeats (IRs) in the chloroplast genomes. The genome regions are colour coded as protein-coding, rRNA-coding, tRNA-coding or conserved non-coding sequences (CNS), whereas arrows indicate the presence of genes

complete chloroplast genome for identification and phylogenetic studies is called super barcoding and has been used successfully in numerous phylogenetic studies (Jansen et al., 2007; Williams, Miller, Small, Nevill, & Boykin, 2016), particularly in resolving taxonomic relationships at lower taxonomic levels (Su et al., 2015). Over the last two decades, the chloroplast genome has been widely used for studying the phylogeny and evolutionary relationships of plants (Jansen et al., 2007; Moore et al., 2007).

A phylogenetic tree was constructed from the data set of 38 cp genomes to understand the *B. sacra* position within the Burseraceae. This phylogenetic tree was conducted based on the complete sequence of the *B. sacra* cp genome, and 37 closely related species were taken from the database for the phylogenetic classification of *B. sacra* within the family Burseraceae and order Sapindales. These phylogenetic analyses were performed with 11 species from Brassicaceae (*Barbarea verna, Pachycladon enysii, Olimarabidopsis pumila, P. cheesemanii, Capsella bursa-pastoris, Arabis hirsuta, Draba nemorosa, Lobularia maritima, Brassica napus, Aethionema cordifolium* and *Aethionema grandiflorum*) and 5 species of

Myrtaceae (*Syzygium cumini*, *Stockwellia quadrifida*, *Eucalyptus umbra*, *E. globulus subsp. globulus* and *E. deglupta*).

The *B. sacra* phylogenetic analysis was conducted using the maximum parsimony method, which resolved 33 nodes for the commercially important species from the Burseraceae family (*C. wightii*) and Sapindales order. The phylogenetic tree is shown with bootstrap support of 55–100% in Fig. 7.4. The *B. sacra* position

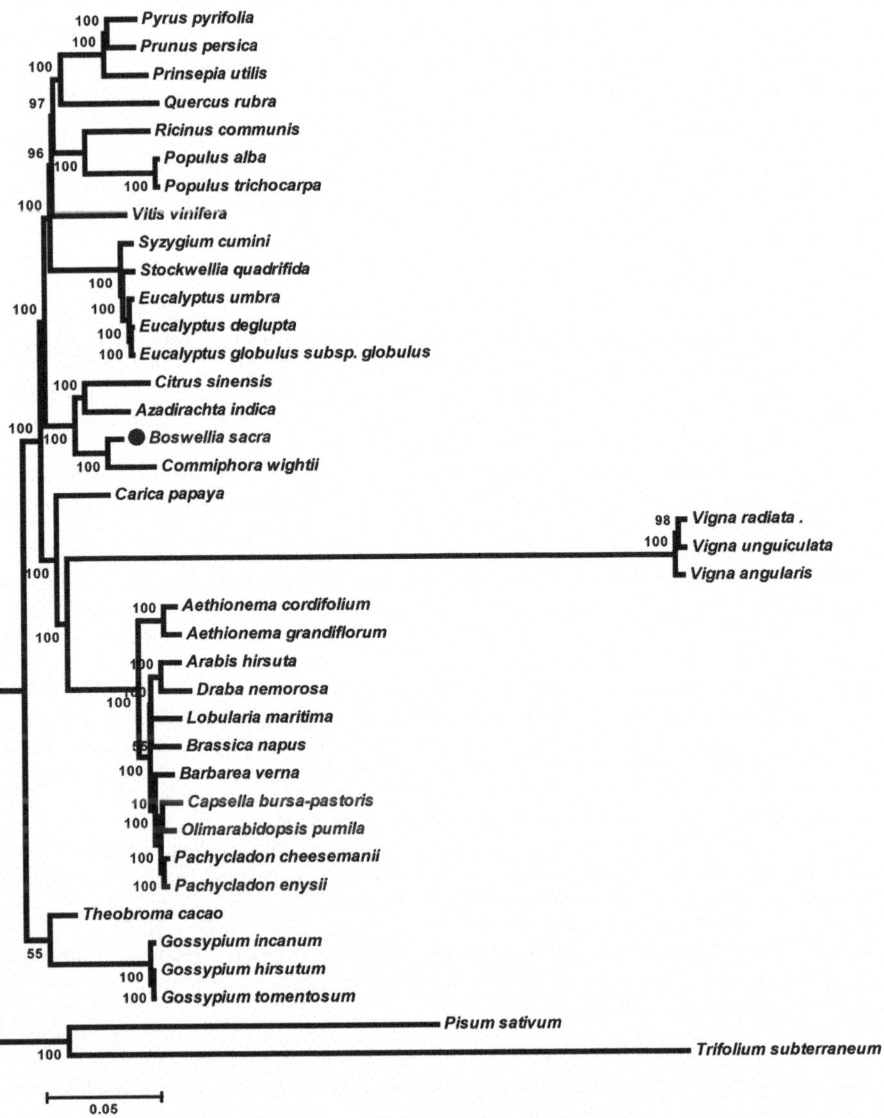

Fig. 7.4 Phylogenetic tree of *Boswellia sacra* cp genome with related species. The whole-genome dataset was analysed using maximum parsimony phylogeny. Numbers above the branches represent bootstrap values. A brown dot represents the position of *B. sacra*

occurs within the Burseraceae family and Sapindales order with high bootstrap support. From the phylogenetic analysis in the present study, it is clear that *B. sacra* and *C. wightii* have a very close relationship and it is supported by high bootstrap support (100%). Previously, Krishnan et al. (Inamdar, Subramanian, & Mohan, 1986) demonstrated the relationship between *A. indica* and *Citrus sinensis*, and they concluded that these species make up a homogenous group (100%), which supports the classification of the *A. indica*. The production of *oleogum* is a common characteristic of *B. sacra, C. wightii* and *A. indica*, while *B. sacra* and *C. wightii* share the same habitat, while the habitats of *A. indica* are different. However, *B. sacra, C. wightii* and *A. indica* belong to the same order, which is an interesting prospect relating to the extensive study in the order Sapindales (Wang, Chen, & Zhang, 2016).

References

Addisalem, A., Esselink, G. D., Bongers, F., & Smulders, M. (2015). Genomic sequencing and microsatellite marker development for *Boswellia papyrifera*, an economically important but threatened tree native to dry tropical forests. *AoB Plants, 7*, plu086.

Asaf, S., Khan, A. L., Khan, M. A., Imran, Q. M., Kang, S.-M., Al-Hosni, K., … Lee, I.-J. (2017). Comparative analysis of complete plastid genomes from wild soybean (*Glycine soja*) and nine other Glycine species. *PLoS One, 12*(8), e0182281.

Corriveau, J. L., & Coleman, A. W. (1988). Rapid screening method to detect potential biparental inheritance of plastid DNA and results for over 200 angiosperm species. *American Journal of Botany, 75*(10), 1443–1458.

Cosner, M. E., Raubeson, L. A., & Jansen, R. K. (2004). Chloroplast DNA rearrangements in Campanulaceae: Phylogenetic utility of highly rearranged genomes. *BMC Evolutionary Biology, 4*(1), 27.

Du, F. K., Lang, T., Lu, S., Wang, Y., Li, J., & Yin, K. (2015). An improved method for chloroplast genome sequencing in non-model forest tree species. *Tree Genetics & Genomes, 11*(6), 114.

Huang, Y.-Y., Matzke, A. J., & Matzke, M. (2013). Complete sequence and comparative analysis of the chloroplast genome of coconut palm (*Cocos nucifera*). *PLoS One, 8*(8), e74736.

Hudson, K. R., & Gardner, R. C. (1988). Organisation of the chloroplast genome of kiwifruit (Actinidia deliciosa). *Current Genetics, 13*(4), 339–342.

Inamdar, J., Subramanian, R. B., & Mohan, J. (1986). Studies on the resin glands of *Azadirachta indica* A. Juss.(Meliaceae). *Annals of Botany, 58*(3), 425–429.

Ivanova, Z., Sablok, G., Daskalova, E., Zahmanova, G., Apostolova, E., Yahubyan, G., & Baev, V. (2017). Chloroplast genome analysis of resurrection tertiary relict *Haberlea rhodopensis* highlights genes important for desiccation stress response. *Frontiers in Plant Science, 8*, 204.

Jansen, R. K., Cai, Z., Raubeson, L. A., Daniell, H., Leebens-Mack, J., Müller, K. F., … Chumley, T. W. (2007). Analysis of 81 genes from 64 plastid genomes resolves relationships in angiosperms and identifies genome-scale evolutionary patterns. *Proceedings of the National Academy of Sciences, 104*(49), 19369–19374.

Jansen, R. K., Kaittanis, C., Saski, C., Lee, S.-B., Tomkins, J., Alverson, A. J., & Daniell, H. (2006). Phylogenetic analyses of Vitis (Vitaceae) based on complete chloroplast genome sequences: Effects of taxon sampling and phylogenetic methods on resolving relationships among rosids. *BMC Evolutionary Biology, 6*(1), 32.

Jansen, R. K., & Ruhlman, T. A. (2012). Plastid genomes of seed plants. In *Genomics of chloroplasts and mitochondria* (pp. 103–126). Dordrecht, The Netherlands: Springer.

Jansen, R. K., Wojciechowski, M. F., Sanniyasi, E., Lee, S.-B., & Daniell, H. (2008). Complete plastid genome sequence of the chickpea (*Cicer arietinum*) and the phylogenetic distribution of rps12 and clpP intron losses among legumes (Leguminosae). *Molecular Phylogenetics and Evolution, 48*(3), 1204–1217.

Jensen, R. G. (2013). Biochemistry of the chloroplast. *Biochemistry Plants, 1*, 273–313.

Jin, S., & Daniell, H. (2015). The engineered chloroplast genome just got smarter. *Trends in Plant Science, 20*(10), 622–640.

Leliaert, F., & Lopez-Bautista, J. M. (2015). The chloroplast genomes of Bryopsis plumosa and Tydemania expeditiones (Bryopsidales, Chlorophyta): Compact genomes and genes of bacterial origin. *BMC Genomics, 16*(1), 204.

Lemieux, C., Otis, C., & Turmel, M. (2014). Six newly sequenced chloroplast genomes from prasinophyte green algae provide insights into the relationships among prasinophyte lineages and the diversity of streamlined genome architecture in picoplanktonic species. *BMC Genomics, 15*(1), 857.

Li, Y., Feng, Y., Wang, X. Y., Liu, B., & Lv, G. H. (2014). Failure of DNA barcoding in discriminating Calligonum species. *Nordic Journal of Botany, 32*(4), 511–517.

Li, Z.-H., Qian, Z.-Q., Liu, Z.-L., Deng, T.-T., Zu, Y.-M., Zhao, P., & Zhao, G.-F. (2016). The complete chloroplast genome of Armand pine *Pinus armandii*, an endemic conifer tree species to China. *Mitochondrial DNA Part A, 27*(4), 2635–2636.

Martin, G., Baurens, F.-C., Cardi, C., Aury, J.-M., & D'Hont, A. (2013). The complete chloroplast genome of banana (*Musa acuminata*, Zingiberales): Insight into plastid monocotyledon evolution. *PLoS One, 8*(6), e67350.

Moore, M. J., Bell, C. D., Soltis, P. S., & Soltis, D. E. (2007). Using plastid genome-scale data to resolve enigmatic relationships among basal angiosperms. *Proceedings of the National Academy of Sciences, 104*(49), 19363–19368.

Moore, M. J., Soltis, P. S., Bell, C. D., Burleigh, J. G., & Soltis, D. E. (2010). Phylogenetic analysis of 83 plastid genes further resolves the early diversification of eudicots. *Proceedings of the National Academy of Sciences, 107*, 4623–4628.

Muse, S. V., & Gaut, B. S. (1994). A likelihood approach for comparing synonymous and nonsynonymous nucleotide substitution rates, with application to the chloroplast genome. *Molecular Biology and Evolution, 11*(5), 715–724.

Nystedt, B., Street, N. R., Wetterbom, A., Zuccolo, A., Lin, Y.-C., Scofield, D. G., … Alexeyenko, A. (2013). The Norway spruce genome sequence and conifer genome evolution. *Nature, 497*(7451), 579.

Ohtani, K., Yamamoto, H., & Akimitsu, K. (2002). Sensitivity to *Alternaria alternata* toxin in citrus because of altered mitochondrial RNA processing. *Proceedings of the National Academy of Sciences, 99*(4), 2439–2444.

Ohyama, K., Fukuzawa, H., Kohchi, T., Shirai, H., Sano, T., Sano, S., … Chang, Z. (1986). Chloroplast gene organization deduced from complete sequence of liverwort *Marchantia polymorpha* chloroplast DNA. *Nature, 322*(6079), 572.

Percy, D. M., Argus, G. W., Cronk, Q. C., Fazekas, A. J., Kesanakurti, P. R., Burgess, K. S., … Graham, S. W. (2014). Understanding the spectacular failure of DNA barcoding in willows (Salix): Does this result from a trans-specific selective sweep? *Molecular Ecology, 23*(19), 4737–4756.

Qian, J., Song, J., Gao, H., Zhu, Y., Xu, J., Pang, X., … Li, C. (2013). The complete chloroplast genome sequence of the medicinal plant Salvia miltiorrhiza. *PLoS One, 8*(2), e57607.

Qin, Z., Wang, Y., Wang, Q., Li, A., Hou, F., & Zhang, L. (2015). Evolution analysis of simple sequence repeats in plant genome. *PLoS One, 10*(12), e0144108.

Raffaelli, M., Mosti, S., & Tardelli, M. (2003). The Frankincense Tree (*Boswellia sacra* Flueck., Burseraceae) in Dhofar, southern Oman: Field-investigations on the natural populations. *Webbia, 58*(1), 133–149.

Saina, J. K., Li, Z.-Z., Gichira, A. W., & Liao, Y.-Y. (2018). The complete chloroplast genome sequence of tree of heaven (*Ailanthus altissima* (Mill.)) (Sapindales: Simaroubaceae), an important Pantropical Tree. *International Journal of Molecular Sciences, 19*(4), 929.

Saski, C., Lee, S.-B., Fjellheim, S., Guda, C., Jansen, R. K., Luo, H., … Clarke, J. L. (2007). Complete chloroplast genome sequences of *Hordeum vulgare*, Sorghum bicolor and Agrostis stolonifera, and comparative analyses with other grass genomes. *Theoretical and Applied Genetics, 115*(4), 571–590.

Seo, T.-K., & Kishino, H. (2008). Synonymous substitutions substantially improve evolutionary inference from highly diverged proteins. *Systematic Biology, 57*(3), 367–377.

Shang, C., Du, F. K., Yin, K., & Zhang, Z. (2016). The complete chloroplast genome of Cathay Poplar: Poplus cathayana Rehder. *Mitochondrial DNA Part B, 1*(1), 86–87.

Shaw, J., Lickey, E. B., Schilling, E. E., & Small, R. L. (2007). Comparison of whole chloroplast genome sequences to choose noncoding regions for phylogenetic studies in angiosperms: The tortoise and the hare III. *American Journal of Botany, 94*(3), 275–288.

Su, H.-J., Hogenhout, S. A., Al-Sadi, A. M., & Kuo, C.-H. (2014). Complete chloroplast genome sequence of Omani lime (*Citrus aurantiifolia*) and comparative analysis within the rosids. *PLoS One, 9*(11), e113049.

Su, J., Zhu, L., Sherman, A., Wang, X., Lin, S., Kamesh, A., … Daniell, H. (2015). Low cost industrial production of coagulation factor IX bioencapsulated in lettuce cells for oral tolerance induction in hemophilia B. *Biomaterials, 70*, 84–93.

Sugiura, M. (1992). The chloroplast genome. *Plant Molecular Biology, 19*(1), 149–168.

Sveinsson, S., & Cronk, Q. (2014). Evolutionary origin of highly repetitive plastid genomes within the clover genus (Trifolium). *BMC Evolutionary Biology, 14*(1), 228.

Wang, L., Wuyun, T.-n., Du, H., Wang, D., & Cao, D. (2016). Complete chloroplast genome sequences of *Eucommia ulmoides*: Genome structure and evolution. *Tree Genetics & Genomes, 12*(1), 12.

Wang, R.-J., Cheng, C.-L., Chang, C.-C., Wu, C.-L., Su, T.-M., & Chaw, S.-M. (2008). Dynamics and evolution of the inverted repeat-large single copy junctions in the chloroplast genomes of monocots. *BMC Evolutionary Biology, 8*(1), 36.

Wang, W.-C., Chen, S.-Y., & Zhang, X.-Z. (2016). Chloroplast genome evolution in Actinidiaceae: clpP loss, heterogenous divergence and phylogenomic practice. *PLoS One, 11*(9), e0162324.

Williams, A. V., Boykin, L. M., Howell, K. A., Nevill, P. G., & Small, I. (2015). The complete sequence of the *Acacia ligulata* chloroplast genome reveals a highly divergent clpP1 gene. *PLoS One, 10*(5), e0125768.

Williams, A. V., Miller, J. T., Small, I., Nevill, P. G., & Boykin, L. M. (2016). Integration of complete chloroplast genome sequences with small amplicon datasets improves phylogenetic resolution in Acacia. *Molecular Phylogenetics and Evolution, 96*, 1–8.

Xie, Q., Michaeli, S., Peled-Zehavi, H., & Galili, G. (2015). Chloroplast degradation: One organelle, multiple degradation pathways. *Trends in Plant Science, 20*(5), 264–265.

Yang, J.-B., Yang, S.-X., Li, H.-T., Yang, J., & Li, D.-Z. (2013). Comparative chloroplast genomes of Camellia species. *PLoS One, 8*(8), e73053.

Yang, Y., Zhou, T., Duan, D., Yang, J., Feng, L., & Zhao, G. (2016). Comparative analysis of the complete chloroplast genomes of five Quercus species. *Frontiers in Plant Science, 7*, 959.

Yao, X., Tang, P., Li, Z., Li, D., Liu, Y., & Huang, H. (2015). The first complete chloroplast genome sequences in Actinidiaceae: Genome structure and comparative analysis. *PLoS One, 10*(6), e0129347.

Yap, J.-Y. S., Rohner, T., Greenfield, A., Van Der Merwe, M., McPherson, H., Glenn, W., … Wilton, A. (2015). Complete chloroplast genome of the wollemi pine (Wollemia nobilis): Structure and evolution. *PLoS One, 10*(6), e0128126.

Zhang, Q., & Liu, Y. (2003). Examination of the cytoplasmic DNA in male reproductive cells to determine the potential for cytoplasmic inheritance in 295 angiosperm species. *Plant and Cell Physiology, 44*(9), 941–951.

Chapter 8
Microbial Communities Accompanying Cultivated and Wild *Boswellia sacra* Trees

Microbial Symbiosis

In this chapter, we investigated the microbial community associated with the rhizospheres of *B. sacra* tree populations. In different *B. sacra* populations, both wild and cultivated trees are always exposed to regenerative threats (Wu et al., 2016). Therefore, rhizosphere microbial communities and their diversity have been reported to play a vital role in promoting plant growth, health and development (Pfeiffer et al., 2016; Taketani et al., 2017). Exploring the role of microbial communities associated with desert plant under extreme conditions provides an opportunity to understand the effect of extreme environments, e.g. desert conditions, on plant growth and development. Microbial symbiosis impacts plant growth and development, making these microbial communities especially interesting to ecologists. Evaluating these communities under marginal conditions provides a new understanding of plant life under water deficient and extreme temperatures. Apart from the unique genetic make-up of desert plants, the unique native microflora associated with plants under these conditions plays a very important role in plant survival. Most of these microbes are predominantly found in the rhizosphere, facilitating plant growth and development by regulating various endogenous processes in plants, e.g. secondary metabolite secretions essential for plants, nutrient assimilation and improving tolerance to various biotic and abiotic stress factors (Andrews & Harris, 2000; Bulgarelli, Schlaeppi, Spaepen, van Themaat, & Schulze-Lefert, 2013; Fonseca-García et al., 2016; Humphrey, Nguyen, Villalobos, & Whiteman, 2014; Lindow & Brandl, 2003; Ritpitakphong et al., 2016). In return, in addition to nutrients and some beneficial exudates, a favourable niche is provided by the plant for the growth and reproduction of specific classes of microbes (Berendsen, Pieterse, & Bakker, 2012; Vandenkoornhuyse, Quaiser, Duhamel, Le Van, & Dufresne, 2015).

Upon associating with plants, both bacteria and fungi provide plants with various types of extracellular enzymes, which interact with various macromolecules

A. Al-Harrasi et al., *Biology of Genus Boswellia*,
https://doi.org/10.1007/978-3-030-16725-7_8

throughout the cells such as proteins, lignin, organic phosphate, carbohydrates and sugar-base polymers to convert them into transportable products (Strong & Claus, 2011; Wingender et al., 2012). The same type of approach has been for observed for a well-known phytohormone, namely, indole acetic acid (IAA), which facilitates root growth and plant development under both stressful and normal environments (Khan et al., 2016). In fact, bacteria produced extracellular enzymes to initiate the host-symbiosis process; however, these extracellular enzymes and IAA increase the plant's tolerance level to both biotic and abiotic stresses by counteracting the pathogenic infections of plants and by regulating various plant tolerance mechanisms (Khan et al., 2016; Ritpitakphong et al., 2016). However, few reports are available about these interactions in desert land plants, specifically *Boswellia sacra* (Knief et al., 2012) (Fig. 8.1).

Microbial Community of *Boswellia sacra*

Very little is known about the relationships of the microbial community associated with the *Boswellia sacra* tree. Most recently, Khan et al. (2016) and El-Nagerabi et al. investigated the microbial community associated with this tree (El-Nagerabi, Elshafie, & Alkhanjari, 2014; Khan et al., 2016). *Boswellia sacra* is unique in terms of its habitats and life cycle in that it experiences a wide range of stresses, e.g. extreme heat, strong wind and drought and UV incidence. The cultivated and wild trees differ greatly in their resin production ability, which is a result of the differences in their growth and development patterns. The microbial flora of important plants (in terms of economics) inhabiting marginalized ecosystems is the least explored area and has been subjected to very few studies, and some of these studies have been restricted to bacterial communities. A few examples

Fig. 8.1 *Boswellia sacra* plants growing in both cultivated and wild conserved areas Khan et al. (2017)

include *Rumex patientia* (Qi, Wang, Xing, Zhao, & Chen, 2012), *Rehmannia glutinosa* (Kuang et al., 2011; Wu et al., 2016), Polygonum cuspidatum (Hao et al., 2012), *Rhododendron arboretum* (Debnath et al., 2016), Aloe vera (Akinsanya, Goh, Lim, & Ting, 2015), *Agave* species (Coleman-Derr et al., 2016), *Thymus zygis* (Pascual et al., 2016) and cacti (Fonseca-García et al., 2016). In this chapter, a detailed bioinformatics analysis and advanced Illumina sequencing were used to explore the four selected plants. The aim of the current study was also to explore the major players in the plant rhizosphere microbiome and identify prominent microbial players.

Diverse Microbiota Associated with *B. sacra* Tree Populations

The fungal and prokaryotic communities of the rhizosphere region in three distinctive *B. sacra* populations, namely, the (i) Adonab (BSA), (ii) Dowkah National Park (BSD) and (iii) Dowkah valley (BSW), were assessed along with comparisons with inclusively grown tree population. The wild populations were grown in water-deficient environments to mimic the desert environment, while cultivated population were provided with the required water and nutrients, and they were grown under the same conditions. The MiSeq sequencing platform was used to elucidate the microbial community's profile. For bacterial and fungal communities, high-quality read data measuring 1.24 Gbp and 678 Mbp, respectively, were produced.

Based on an OTU analysis, the bacteria exhibited significantly higher ($P < 0.0019$; average 1006) OTUs compared to the fungi, which had significantly lower results ($P < 0.0019$; average 60.6), hence suggesting the enormous microbial diversity of the bacterial species. OTU analyses were also performed for population-specific fungi, suggesting that BSD and BSW showed significantly lower OTUs compared to those of BSA, which were high ($P < 0.02$; 102). However, the Shannon analysis showed a significantly higher value in BSD (ANOSIM; $R^2 = 0.718$, $p < 0.00071$) compared to the BSW and BSA populations (Fig. 8.2). Among bacterial communities, the population was noted to have significantly higher OTUs ($P < 0.001$; 1178) in comparison to the BSD and BSW. The same type of result was recorded for Shannon diversity indices (Fig. 8.2). Based on the OTUs and rate of abundance, the bacterial communities had the highest indices compared to the fungal communities associated with *B. sacra* trees (Fig. 8.2). All the fungal populations were significantly different from one another based on the NMDS analyses of fungal communities. Nonetheless, the BSW and BSD populations were very similar ($R^2 = 0.78$; $P < 0.031$, respectively, Fig. 8.3). However, BSW was found to make up a separate group based on the NMDS analysis of the bacterial communities ($R^2 = 0.937$; $P < 0.0092$). This result suggests no difference in the replicates from BSW, and the BSA replicates were significantly different from one another, suggesting the heterogeneous diversity of fungal OTUs among replicates. BSD showed a smaller distance among the replicates compared to BSA. The BSD replicates were not distantly separated from one another.

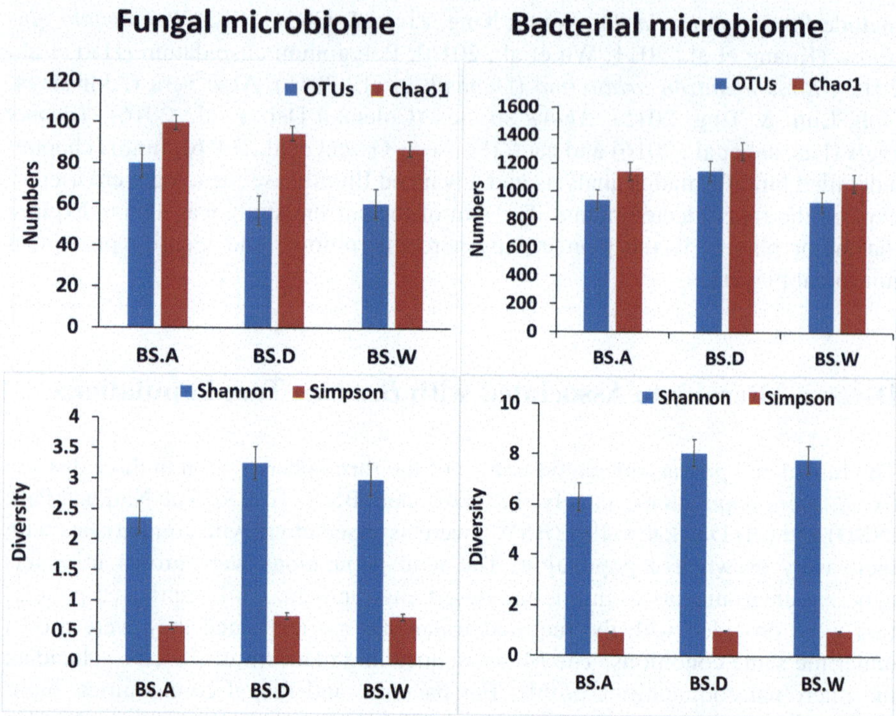

Fig. 8.2 Bacterial and fungal community distribution of Chao-1 and operational taxonomic units (OTUs) of replicates from the rhizosphere samples from both wild and cultivated *B. sacra* trees according to ITS and 16S sequencing through MiSeq Khan et al. (2017)

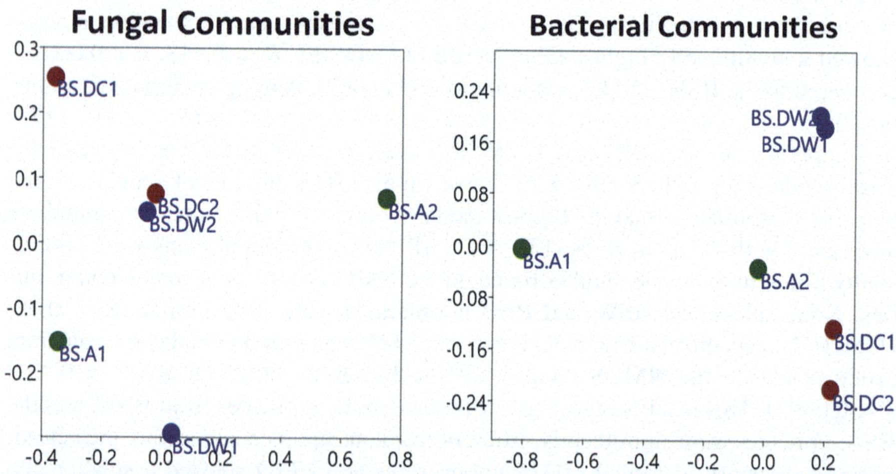

Fig. 8.3 Nonmetric multidimensional scaling (NMDS) plots for Bray-Curtis distances of different rhizosphere populations in *B. sacra* Khan et al. (2017)

B. sacra Rhizosphere Fungal Microbiota

Among the three populations, the rhizosphere microbial diversity was found to be very distant from one another. Among the different fungal phyla, *Basidiomycota*, *Ascomycota*, *Chytridiomycota*, *Zygomycota* and *Glomeromycota* were present in high numbers in all three populations (BSA, BSD and BSW). In the wild population (BSW), *Ascomycota* was found at significantly higher concentrations ($P < 0.001$; 60%) followed by BSA (47%) *Basidiomycota*, while on the other side, the *Ascomycota* was found to be significantly higher ($P < 0.001$; 52%) in the BSA rhizosphere (Fig. 8.4). A major proportion of shared data was noted among the unidentified fungal communities, which made up approximately 76% and 65% of the BSW and BSD populations, respectively. *Zygomycota* and *Glomeromycota* were abundant in BSD.

Regarding the phyla correlation, in the *B. sacra* population, number two were distributed throughout all three *B. sacra populations* (Fig. 8.4). *Haematonectria*

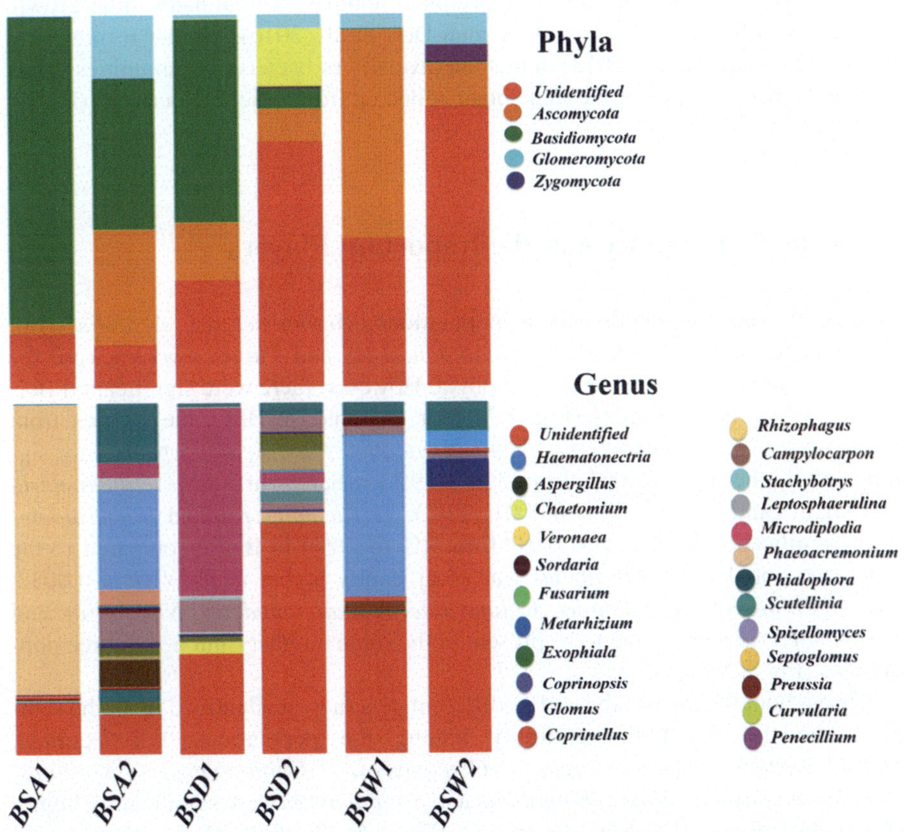

Fig. 8.4 Fungal phyla and genera relative abundances and shared core OTUs in cultivated and wild *B. sacra* populations Khan et al. (2017)

was abundant in all three populations (1.1–8.1%). In BSA, the concentrations of the genera *Coprinopsis*, *Exophiala*, *Veronaea* and *Aspergillus* were significantly higher ($P < 0.001$; 4–7.6%) compared to the BSW and BSD samples. *Rhizophagus* and *Glomus* were exclusively rich in the BSD population. Notably, the genera *Spizellomyces* and *Chaetomium* were highly abundant in BSW ($P < 0.0001$; 4–7.8%) compared with the other two samples (Fig. 8.4).

The results revealed the bacterial and fungal communities at the three different locations, and there were different responses in the metagenomic data output as well as the OTU numbers. Furthermore, various researchers reported that some host-specific characteristics and specific genetic pathways can have vital effects on the microbial composition and diversity (Kembel et al., 2011; Horton et al., 2014; Lebeis et al., 2015; Van and Dufresne, 2015). Although the soil and climatic conditions of these populations are very similar, due to regular irrigation, the BSW populations are different from the BSD and BSA populations. Therefore, the wild and cultivated plants and their rhizosphere have revealed a different niche of microbiota. Similar results were reported previously by various researchers who uncovered significant differences among various agave plant populations that had similar growth conditions to those of *B. sacra* (Coleman-Derr et al., 2016). Similar results were reported by Peiffer et al. (2013) in that the diversity of bacterial communities varies among the rhizospheres of the same plant collected from different locations (Peiffer et al., 2013).

Bacterial Community and the Important Players

In the rhizosphere of the three populations, *Actinobacteria*, *Acidobacteria*, *Proteobacteria*, *Bacteroidetes*, *Gemmatimonadetes* and *Planctomycetes* were the most abundant ($P < 0.0001$) bacterial phyla. However, there were significant differences among these samples (Fig. 8.5). For example, in BSD, the species from *Planctomycetes*, *Proteobacteria*, *Gemmatimonadetes* and *Acidobacteria* were significantly higher ($P < 0.0001$; from 5% to 29%), whereas in BSW, *Actinobacteria* was significantly higher ($P < 0.0001$; 27%). *Verrucomicrobia* and *Bacteroidetes* were significantly higher ($P < 0.01$; from 4% to 28%) in BSA. Compared to the other two populations, *Firmicutes* was significantly higher in BSW ($P < 0.0032$; 5.2%). *Chloroflexi*, *Candidatus*, *Chlamydiae*, *Armatimonadetes*, *Nitrospirae* and *Latescibacteria* were found to be present in the same numbers among all three populations (Fig. 8.5).

Throughout the data analysis, 191 different genera were identified in all the samples, showing the high variability among the populations. For example, *Steroidobacter*, *Ohtaekwangia*, *Glycomyces*, *Chitinophaga*, *Opitutus*, *Promicromonospora*, *Novosphingobium* and *Streptomyces* were significantly higher ($P < 0.0001$; from 1.2% to 20.1%) in BSA (Fig. 8.5). *Pirellula*, *Gemmatimonas* and *Lysobacter* were significantly abundant ($P < 0.0001$; from 1% to 6%) in BSD rhizosphere samples. However, *Planctomyces*, *Phenylobacterium* and *Sphingomonas*

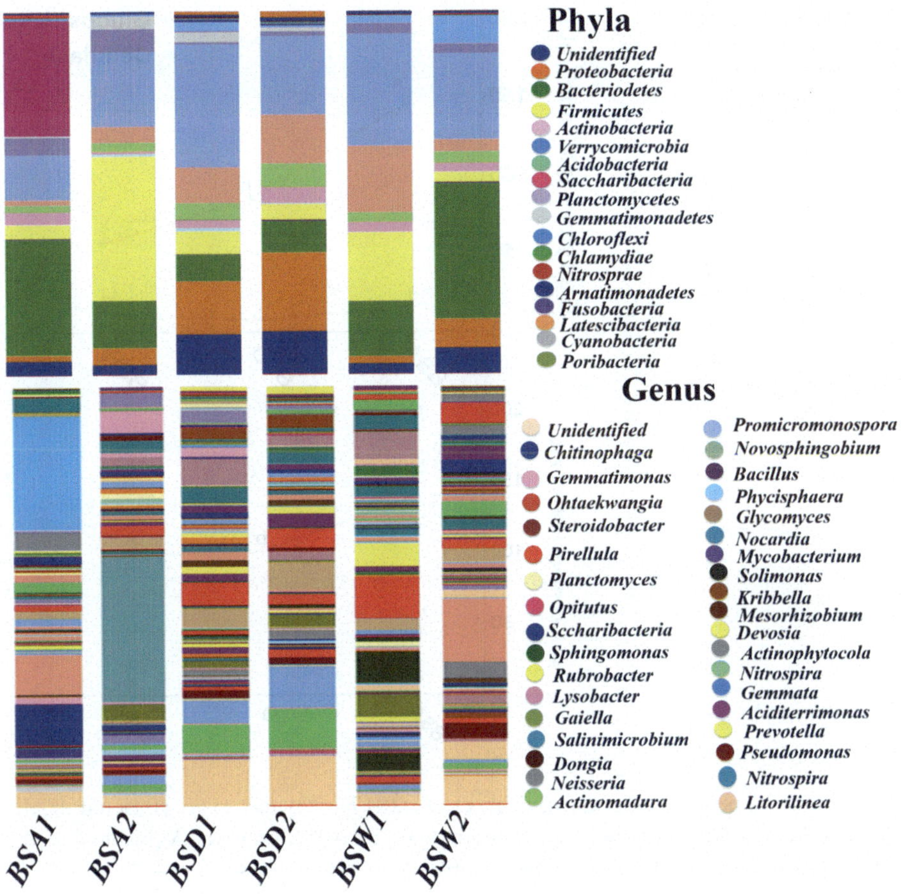

Fig. 8.5 Bacterial phyla and genera abundances and shared core OTUs in the rhizosphere soil of cultivated and wild *B. sacra* populations Khan et al. (2017)

were significantly abundant in BSW ($P < 0.008$; from 1% to 8.6%) (Fig. 8.5). *Mycobacterium*, *Gemmatimonas*, *Gaiella*, *Ohtaekwangia* and *Dongia* were found in the BSA, BSD and BSW samples (Fig. 8.6). For bacterial species, uncultured bacteria were abundantly present in all the samples. However, *Acidobacteria* sp., *Actinomycete* sp., *Planctomycete* sp., *Gammaproteobacterium*, *Firmicutes* sp., *Prevotella nanceiensis*, *Escherichia coli*, *Chloroflexi* sp., *Lactobacillus reuteri*, *Rubrobacter* sp., *Actinoplanes* sp. and *Bacillus niabensis* were identified and found across the rhizospheres of BSA, BSD and BSW, ranging from 0.5% to 6.28%.

The results revealed that in the bacterial communities, *Proteobacteria*, *Acidobacteria*, *Bacteroidetes* and *Actinobacteria* were the significant bacterial phyla from the three rhizosphere populations. Similar results were reported by Akinsanya et al. (2015) and Garcia et al. (2016) in which the above species were the most abundant species in various plants. Similar results were also reported in various

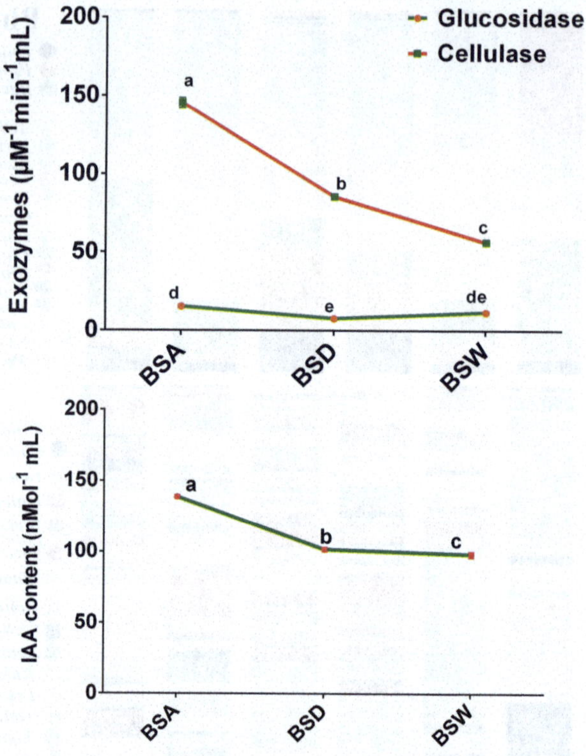

Fig. 8.6 Indole acetic acid and exozyme (cellulases and glucosidases) quantification in cultivated and wild *B. sacra* rhizospheres

medicinal plants such as ginseng (Tan et al., 2017), *Thymus zygis* (Pascual et al., 2016), *Rhododendron arboreum* (Debnath et al., 2016), Aloe vera (Akinsanya et al., 2015) (Golinska et al. 2015), etc. The occurrence of core phyla such as *Actinobacteria* and *Proteobacteria* in BW revealed the higher potential of the host plant to cope with harsh environmental conditions. Similar results were also reported by Marasco et al. (Marasco et al., 2012) and Kaplan et al. (Kaplan et al., 2013) in that higher abundances of *Actinobacteria* and *Proteobacteria* will help the plant under harsh environmental conditions.

IAA and Exozyme Production in the *B. sacra* Rhizosphere

The glucosidases and cellulases were conducted and quantified in the rhizosphere samples from the BSA, BSW and BSD populations. Cellulases were found to be significantly higher in BSA rhizosphere samples ($P < 0.0001$; 148.4 ± 1.8 µM/min/mL), followed by BSD (91.8 ± 2.3 µM/min/mL) (Fig. 8.6). Lower amounts of cellulases were detected in the rhizosphere of wild *B. sacra*. Among all the rhizosphere samples, low concentrations were recorded for glucosidases. However, BSA had a significantly higher ($P < 0.018$; 17.1 ± 0.9 µM/min/mL) amount of glucosidases compared to BSW and BSD. The IAA content was significantly higher in BSA ($P < 0.001$; 137.8 ± 2.1 nM/mL) compared to BSW and BSD (Fig. 8.6).

The results provide a genomic baseline to investigate the interactions of complex microbial communities with the *B. sacra* plant in arid regions, even though very limited material is available for studies on *B. sacra*. Furthermore, the identification and characterization of various taxa can provide new insights for future research related to *B. sacra* about how these rhizosphere microbes play significant roles in plant growth and development.

References

Akinsanya, M. A., Goh, J. K., Lim, S. P., & Ting, A. S. Y. (2015). Metagenomics study of endophytic bacteria in *Aloe vera* using next-generation technology. *Genom Data, 6*, 159–163.

Andrews, J. H., & Harris, R. F. (2000). The ecology and biogeography of microorganisms on plant surfaces. *Annual Review of Phytopathology, 38*(1), 145–180.

Berendsen, R. L., Pieterse, C. M., & Bakker, P. A. (2012). The rhizosphere microbiome and plant health. *Trends in Plant Science, 17*(8), 478–486.

Bulgarelli, D., Schlaeppi, K., Spaepen, S., van Themaat, E. V. L., & Schulze-Lefert, P. (2013). Structure and functions of the bacterial microbiota of plants. *Annual Review of Plant Biology, 64*, 807–838.

Coleman-Derr, D., Desgarennes, D., Fonseca-Garcia, C., Gross, S., Clingenpeel, S., Woyke, T., … Tringe, S. G. (2016). Plant compartment and biogeography affect microbiome composition in cultivated and native Agave species. *New Phytologist, 209*(2), 798–811.

Debnath, R., Yadav, A., Gupta, V. K., Singh, B. P., Handique, P. J., & Saikia, R. (2016). Rhizospheric bacterial community of endemic Rhododendron arboreum Sm. Ssp. delavayi along eastern Himalayan slope in Tawang. *Frontiers in Plant Science, 7*, 1345.

El-Nagerabi, S. A., Elshafie, A. E., & Alkhanjari, S. S. (2014). Endophytic fungi associated with endogenous *Boswellia sacra*. *Biodiversitas Journal of Biological Diversity, 15*(1).

Fonseca-García, C., Coleman-Derr, D., Garrido, E., Visel, A., Tringe, S. G., & Partida-Martínez, L. P. (2016). The cacti microbiome: Interplay between habitat-filtering and host-specificity. *Frontiers in Microbiology, 7*, 150.

Garcia, A., Polonio, J., Polli, A., Santos, C., Rhoden, S., Quecine, M., … Pamphile, J. (2016). Rhizosphere bacteriome of the medicinal plant *Sapindus saponaria* L. revealed by pyrosequencing. *Genetics and molecular research: GMR, 15*(4).

Golinska, P., Wypij, M., Agarkar, G., Rathod, D., Dahm, H., & Rai, M. (2015). Endophytic actinobacteria of medicinal plants: diversity and bioactivity. *Antonie Van Leeuwenhoek, 108*(2), 267–289.

Hao, D., Ma, P., Mu, J., Chen, S., Xiao, P., Peng, Y., … Sun, C. (2012). De novo characterization of the root transcriptome of a traditional Chinese medicinal plant *Polygonum cuspidatum*. *Science China Life Sciences, 55*(5), 452–466.

Horton, M. W., Bodenhausen, N., Beilsmith, K., Meng, D., Muegge, B. D., Subramanian, S., ... & Bergelson, J. (2014). Genome-wide association study of Arabidopsis thaliana leaf microbial community. *Nature Communications, 5*, 5320.

Humphrey, P. T., Nguyen, T. T., Villalobos, M. M., & Whiteman, N. K. (2014). Diversity and abundance of phyllosphere bacteria are linked to insect herbivory. *Molecular Ecology, 23*(6), 1497–1515.

Kaplan, D., Maymon, M., Agapakis, C. M., Lee, A., Wang, A., Prigge, B. A., … Hirsch, A. M. (2013). A survey of the microbial community in the rhizosphere of two dominant shrubs of the Negev Desert highlands, Zygophyllum dumosum (Zygophyllaceae) and *Atriplex halimus* (Amaranthaceae), using cultivation-dependent and cultivation-independent methods. *American Journal of Botany, 100*(9), 1713–1725.

Kembel, S. W., Eisen, J. A., Pollard, K. S., & Green, J. L. (2011). The phylogenetic diversity of metagenomes. *PLoS One 6*: e23214.

Khan, A. L., Al-Harrasi, A., Al-Rawahi, A., Al-Farsi, Z., Al-Mamari, A., Waqas, M., … Shin, J.-H. (2016). Endophytic fungi from frankincense tree improves host growth and produces extracellular enzymes and indole acetic acid. *PLoS One, 11*(6), e0158207.

Khan, A. L., Asaf, S., Al-Rawahi, A., Lee, I. J., & Al-Harrasi, A. (2017). Rhizospheric microbial communities associated with wild and cultivated frankincense producing Boswellia sacra tree. *PloS one, 12*(10), e0186939.

Knief, C., Delmotte, N., Chaffron, S., Stark, M., Innerebner, G., Wassmann, R., … Vorholt, J. A. (2012). Metaproteogenomic analysis of microbial communities in the phyllosphere and rhizosphere of rice. *The ISME Journal, 6*(7), 1378.

Kuang, D.-Y., Wu, H., Wang, Y.-L., Gao, L.-M., Zhang, S.-Z., & Lu, L. (2011). Complete chloroplast genome sequence of Magnolia kwangsiensis (Magnoliaceae): Implication for DNA barcoding and population genetics. *Genome, 54*(8), 663–673.

Lebeis, S. L., Paredes, S. H., Lundberg, D. S., Breakfield, N., Gehring, J., McDonald, M., ... & Dangl, J. L. (2015). Salicylic acid modulates colonization of the root microbiome by specific bacterial taxa. *Science, 349*(6250), 860–864.

Lindow, S. E., & Brandl, M. T. (2003). Microbiology of the phyllosphere. *Applied and Environmental Microbiology, 69*(4), 1875–1883.

Marasco, R., Rolli, E., Ettoumi, B., Vigani, G., Mapelli, F., Borin, S., … Cherif, A. (2012). A drought resistance-promoting microbiome is selected by root system under desert farming. *PLoS One, 7*(10), e48479.

Pascual, J., Blanco, S., García-López, M., García-Salamanca, A., Bursakov, S. A., Genilloud, O., … van Dillewijn, P. (2016). Assessing bacterial diversity in the rhizosphere of Thymus zygis growing in the Sierra Nevada National Park (Spain) through culture-dependent and independent approaches. *PLoS One, 11*(1), e0146558.

Peiffer, J. A., Spor, A., Koren, O., Jin, Z., Tringe, S. G., Dangl, J. L., … Ley, R. E. (2013). Diversity and heritability of the maize rhizosphere microbiome under field conditions. *Proceedings of the National Academy of Sciences, 201302837.*

Pfeiffer, S., Mitter, B., Oswald, A., Schloter-Hai, B., Schloter, M., Declerck, S., & Sessitsch, A. (2016). Rhizosphere microbiomes of potato cultivated in the high Andes show stable and dynamic core microbiomes with different responses to plant development. *FEMS Microbiology Ecology, 93*(2), fiw242.

Qi, X., Wang, E., Xing, M., Zhao, W., & Chen, X. (2012). Rhizosphere and non-rhizosphere bacterial community composition of the wild medicinal plant *Rumex patientia. World Journal of Microbiology and Biotechnology, 28*(5), 2257–2265.

Ritpitakphong, U., Falquet, L., Vimoltust, A., Berger, A., Métraux, J. P., & L'Haridon, F. (2016). The microbiome of the leaf surface of Arabidopsis protects against a fungal pathogen. *New Phytologist, 210*(3), 1033–1043.

Strong, P., & Claus, H. (2011). Laccase: A review of its past and its future in bioremediation. *Critical Reviews in Environmental Science and Technology, 41*(4), 373–434.

Taketani, R. G., Lançoni, M. D., Kavamura, V. N., Durrer, A., Andreote, F. D., & Melo, I. S. (2017). Dry season constrains bacterial phylogenetic diversity in a semi-arid rhizosphere system. *Microbial Ecology, 73*(1), 153–161.

Tan, Y., Cui, Y., Li, H., Kuang, A., Li, X., Wei, Y., & Ji, X. (2017). Diversity and composition of rhizospheric soil and root endogenous bacteria in Panax notoginseng during continuous cropping practices. *Journal of Basic Microbiology, 57*(4), 337–344.

Vandenkoornhuyse, P., Quaiser, A., Duhamel, M., Le Van, A., & Dufresne, A. (2015). The importance of the microbiome of the plant holobiont. *New Phytologist, 206*(4), 1196–1206.

Wingender, G., Stepniak, D., Krebs, P., Lin, L., McBride, S., Wei, B., … Kronenberg, M. (2012). Intestinal microbes affect phenotypes and functions of invariant natural killer T cells in mice. *Gastroenterology, 143*(2), 418–428.

Wu, P., Xiong, X., Xu, Z., Lu, C., Cheng, H., Lyu, X., … Lyu, Y. (2016). Bacterial communities in the rhizospheres of three mangrove tree species from Beilun Estuary, China. *PLoS One, 11*(10), e0164082.

Chapter 9
Endophytic Microbial Communities of *Boswellia*

Endophytic Microbes

De Bary (1879) introduced symbiosis as one of the most essential associations of two or more living organisms (De Bary, 1879). Each partner receives several benefits during symbiotic interactions (Lewis, 1985). In fungal host-plant symbiotic associations, the consequences for both partners may be positive (commensalism, mutualism or parasitism), negative (competition) or neutral (neutralism and amensalism) (Rodriguez et al., 2008). Compatible fungal host-plant interactions involve at least three events, and the symbionts as well as the pathogens must triumph over the host's surveillance system (Krings et al., 2007). The sequential events consist of fungal hyphae penetration into plant tissues, the invasion of the plant tissues by colonization and the expression of a symbiotic lifestyle. During the establishment of symbiotic associations, biochemical and/or genetic communication is thought to occur between the endophytes, and the invasion of the host plant allows for the exchange, conferring physiological benefits to each organism.

Endophytes are ***pervasive*** microorganisms (both fungi and bacteria) that reside in tissues of plants while not causing any diseases. The term 'endophyte' is applied to fungal or bacterial microorganisms that colonize host-plant organs such as the roots, stems, leaves and even seeds, and they spend all or part of their life cycle in the host and cause no apparent injury (Saikkonen, Wäli, Helander, & Faeth, 2004; Wennstrom, 1994; Wilson, 1993). These endophytic microorganisms are associated with the plant throughout their life history, starting from seed germination to fruit development. Endophytic microorganisms have been reported to be found in the rhizosphere (roots), phylloplane (leaves), laimosphere/caulosphere (stem), carposphere (fruit), spermosphere (seed) and anthosphere (flower) (Clay, 1990; Lindow & Brandl, 2003; Saikkonen et al., 2004; Sessitsch et al., 2012). Most of the vascular and non-vascular plants species studied in natural ecosystems harbour endophytic fungi that increase the physiological robustness of their host plants (Müller & Krauss, 2005). This broad term explicitly excludes the functional mycorrhizal fungi;

A. Al-Harrasi et al., *Biology of Genus Boswellia*,
https://doi.org/10.1007/978-3-030-16725-7_9

however, it does not suggest that endophytes cannot be cultured in microbial growth media. Most of the definitions indicate that during *biological association* with the living host plants, endophytic fungi have to exhibit a mycelial lifestyle for at least some time. Therefore, to recognize a fungus as an 'endophyte', the observer should note the presence of its *hyphae in living tissues* (Kuldau & Bacon, 2008; Maheshwari, 2006). Endophytic fungi are highly diverse in nature; they occur within asymptomatic tissues of non-vascular plants and seed plants from the tundra to hot deserts and from crop plants to the most biodiverse tropical forests and mangroves (Arnold, Henk, Eells, Lutzoni, & Vilgalys, 2007). Current investigations suggest that symbiotic associations significantly influence host-plant characteristics through the existence of fungi and other microorganisms in the internal tissues and they occupy the leaf and root surfaces and the rhizosphere (Ganley, Brunsfeld, & Newcombe, 2004; Van Der Heijden, Bardgett, & Van Straalen, 2008). Furthermore, it is now well known that symbiosis is unavoidable for some plants to survive in nature (Sapp, 2004). Fungal symbiosis is the key factor for starting a successful life on the earth's surface. The record of this fungal symbiotic association traces back to at least 400 million years ago as revealed by the fossil record (Krings et al., 2007; Pirozynski & Malloch, 1975; Redecker, Kodner, & Graham, 2000).

During symbiosis, endophytes follow different strategies for establishing associations and disturbing plant defence signalling networks or reprogramming the host metabolism by modifying hormonal homoeostasis and antioxidant contents (Göhre & Robatzek, 2008; Robert-Seilaniantz, Navarro, Bari, & Jones, 2007). Apart from those roles, although endophytic fungi are hidden, they display high fungal biodiversity. The estimated number of fungal species is 1.5 million; however, only one million or fewer have been described (Fröhlich & Hyde, 1999; Hawksworth & Rossman, 1997). One example can clearly explain the remaining undiscovered species of fungi. Hibbett et al. (2011) suggested that this level of diversity could include more than 8.7 million species in both the rhizosphere and the phyllosphere (Hibbett et al., 2011). Perhaps every plant species encompasses a wide range of endophytes, e.g. 90 species in a leaf (tropical tree), 50 genera from the roots of arid land species and desert plants encompass less than 20 species in leaves (Bayman & Otero, 2006; Khan, Waqas, et al., 2017; Porras-Alfaro et al., 2008). The colonization rates of endophytes vary in different plant species from <1% to 41% in the leaf part from boreal/arctic ecosystems up to 90% in tropical forest ecosystems (Higgins, Arnold, Miadlikowska, Sarvate, & Lutzoni, 2007; Lodge, Fisher, & Sutton, 1996). Endophytes residing in a single plant while living in various tissues parts are either interacting with one another or they are not dependant on the presence of others (Khan et al., 2012; Stone, Bacon, & White Jr, 2000). The unrevealed biodiversity and issue of host specificity with special reference to tropical plants have increased the interest of mycologists in this subject.

Endophytic Microorganisms and Tree Growth

The production of phytohormone-like chemical constituents is considered essential for plant growth-promoting effects. Endophytes are also observed for plant hormones. The plant growth-promoting hormone indole-3-acetic acid (IAA) was reisolated from the tall fescue fungal endophytes *Acremonium coenophialum* (De Battista, Bacon, Severson, Plattner, & Bouton, 1990; Rowan, Latch, Bacon, & White, 1994), *Epicoccum purpurascens*, *Colletotrichum* sp. and *Aureobasidium pullulans* (Petrini, Sieber, Toti, & Viret, 1993). Along with indole acetic acid (IAA) and indole acetonitrile, cytokinin was also detected in the tobacco-isolated endophyte *Hypoxylon serpens* (Dew et al., 1990; Petrini et al., 1993). Various bioactive secondary metabolites obtained from endophytes are utilized as potent drugs against several diseases (Kharwar, Mishra, Gond, Stierle, & Stierle, 2011; Kusari, Hertweck, & Spiteller, 2012). The most important example is the anticancer compound Taxol, which was first reported as the product of fungal endophytes (*Taxomyces andreanae*). This endophyte was isolated from *T. brevifolia* (Stierle, Strobel, & Stierle, 1993). Another example of an anticancer drug from an endophyte is camptothecin and the structural analogues hypericin and azadirachtin (Kusari & Spiteller, 2012; Kusari, Zühlke, & Spiteller, 2009).

In addition to the production of plant growth-promoting hormones and essential secondary metabolites, endophytic microbes were recently found to produce a wide array of exozymes (extracellular enzymes), e.g. phosphatase, glucosidase, cellulase, xylanase, esterases, etc. (Corrêa et al., 2014; Esteves, Saraiva, Correia, & Alves, 2014; Khan, Waqas, et al., 2015; Voříšková & Baldrian, 2013). The production of these types of exozymes is essential to the host plant and will not require additional metabolism under stress conditions, and thus they could also act as plant energy savers. This is also true for the production of phytohormone-like compounds (Schulz & Boyle, 2005; Soussi et al., 2016). Exploring symbiotic endophytic microorganisms associated with economically important and unique endemic plant species is essential for understanding plant life as well as providing an alternative source for plant-like metabolites.

Endophytic Microbes from Frankincense Trees

The frankincense tree grows in subtropical and extremely arid environments. The amount of water and nutrients available to the tree vary greatly in the wild. In addition to physiochemical stress, the *Boswellia* species are prone to herbivory by insects, camels and rodents and to fungal attacks following resin tapping. These extreme conditions often provide synergistic and friendly conditions for plants to allow microbes to associate with them quickly (Kim et al., 2018). Since *B. sacra* is very difficult to propagate due to its harsh nature, using endemic microflora as a plant growth promoter can help ameliorate the decline in the tree population

Fig. 9.1 *B. sacra* tree species and a method used to collect different tissues from different parts of the tree to isolate and identify endophytic microbes (both bacteria and fungi). (Source: Khan, Al-Harrasi, et al., 2016)

[Fig. 9.1; (Hibbett et al., 2011)]. Isolating and identifying these novel endophytes can later provide endemic inoculum to enhance plant propagation strategies. The resin-based chemical profile of *Boswellia* has potential benefits in treating human diseases and has been extensively elucidated in the recent literature as an analgesic (Tolera, Sass-Klaassen, Eshete, Bongers, & Sterck, 2013), hepatoprotective (Schulz & Boyle, 2005), antioxidant (Stierle et al., 1993), anticoagulant (Gunatilaka, 2006), tumour suppressor (Khan & Lee, 2013), anti-inflammatory (Khan et al., 2012) and cardioprotective agent (Khan, Hussain, Al-Harrasi, Al-Rawahi, & Lee, 2015), and it has also been used in the treatment of gastric, hepatic and skin disorders (Limtong, Kaewwichian, Yongmanitchai, & Kawasaki, 2014; Petrini et al., 1993). However, the potential world of endophytic microorganisms has not been fully elucidated until now, which could also lead to a future research project as begun by Khan et al. (2016) (Khan, Al-Harrasi, et al., 2016; Khan et al., 2018); (Figs. 9.1 and 9.2).

In a recent study by Khan et al. (2016), diverse endophytic fungi belonging to various classes were isolated from stem and leaf parts of the tree (Fig. 9.2).

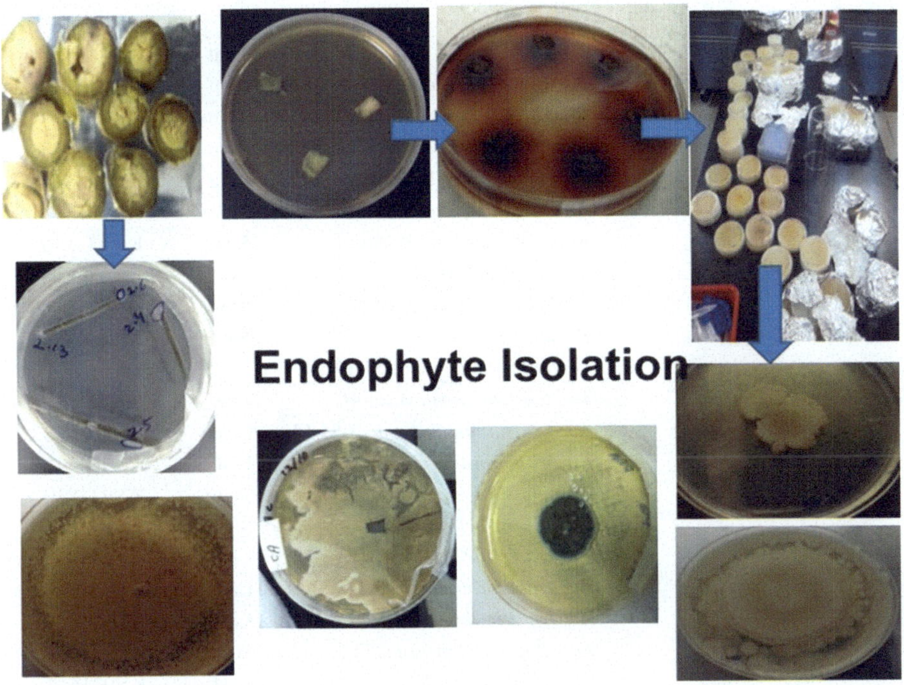

Fig. 9.2 Isolated fungal endophytes. Endophytes from different phyllospheric parts of the frankincense tree

The results led to the identification of 32 species of endophytes (Table 9.1; Figs. 9.3 and 9.4). The authors suggested that the colonization pattern was significantly higher in leaves than in the stem parts. The different diversity indices such as the Shannon-Weiner diversity index and the Simpson index recorded 0.8729 and 0.583, respectively, for leaf parts compared to stem parts. The endophytic microbes were identified through molecular and sequencing approaches (Figs. 9.3 and 9.4).

For the morphological analysis, different parameters such as the colour of the aerial hyphae, surface texture, colony shape, height from the medium, base colour, growth rate and pattern were observed as suggested by Arnold et al. (2007) (Arnold & Lutzoni, 2007). In another study by (Ali, Charles, & Glick, 2012), the endophyte isolations were analysed by Shannon-Weiner indices ($H1 = 0.907$) and Simpson indices ($1/l$=1.77). These endophytic fungi were further identified by extracting the gDNA, amplifying it by PCR, sequencing the 18S rRNA region, analysing the product with bioinformatics tools (BLAST search) and performing a detailed phylogenetic analysis. A conclusion showing higher endophytic abundance in the leafy part of the tree was previously suggested by Yan et al. (Yan, Broughton, Yang, & Gange, 2015), Sakayaroj et al. (Sun et al., 2014), El-Nagerabi et al. (El-Nagerabi, Elshafie, & Alkhanjari, 2014), Ghimire et al. (Ghimire, Charlton, Bell, Krishnamurthy, & Craven, 2011), Santamaria and Bayman (Santamaría & Bayman, 2005; Sun, Wang,

Table 9.1 Endophytic fungal strains isolated from the stem and leaf parts of the frankincense tree

Name	Tree part	Accession no
Preussia sp. BSL10	Leaf	KR231682
Penicillium citrinum BSL21	Leaf	KX233837
Phoma medicaginis BSS14	Stem	KR231679
Aureobasidium pullulans BSS1-2b	Stem	KX233835
Chaetomium sp. BSL11	Leaf	KX233834
Thielavia arenaria BSS7	Stem	KR231678
Aureobasidium sp. BSS6	Stem	KX233836
Preussia sp. BSL12	Leaf	KR231680
Dothideomycetes sp. BSS11	Stem	KR231681
Phoma sp. BSS1	Stem	KR231683
Penicillium citrinum BSL17	Leaf	KR231677
Thielavia arenaria BSPDB	Leaf	KX233840
Aureobasidium pullulans BSL13	Leaf	KX233838
Fusarium proliferatum B3	Leaf	KX233842
Aureobasidium pullulans BSS15	Stem	KX233839
Sordariomycetes sp. L2	Stem	KX233841
Preussia sp. BSL2-1	Leaf	KX233846
Fusarium proliferatum LF3	Leaf	KY921610
Aspergillus nidulans BSL2-1	Leaf	KY921609
Paecilomyces variotii SF2	Stem	KY921608
Alternaria sp. RF-6B	Leaf	KY921613
Epicucum nigram SF3	Stem	KY921611
Penicillium purpurogenum SF3	Stem	KY921612

Lu, Okane, & Kakishima, 2011), Garcia et al. (García, Rhoden, Rubin Filho, Nakamura, & Pamphile, 2012), Sun et al. (Sun et al., 2014), Liu et al. (Liu et al., 2007), Kumar and Hyde (Kumar & Hyde, 2004) and Huang et al. (El-Nagerabi et al., 2014). Similarly, a study by El-Nagerabi et al. (Arenal, Platas, & Pelaez, 2007) showed that 61% of endophytic microbes are present on the leaf part of *B. sacra*. In the current literature, we calculate that the total number of reported endophytic species from the frankincense tree (*B. sacra*) is approximately 121 strains. However, this number could be even higher. Further endophytic microbiome analysis through next-generation sequencing would certainly help to uncover more interesting information.

Methods for Isolating Endophytes from *Boswellia* Trees

Some recent studies have shown that various tree parts, especially the stem and leaf parts of *Boswellia*, were used to isolate endophytic microbes (Figs. 9.1 and 9.2). However, no research report has shown that the root parts were considered to contain an endophytic microbiome, although a recent study by Khan et al. (2017) was performed on microbial communities of the rhizosphere region. A schematic representation of how the endophytic microbes were isolated is shown in Fig. 9.5.

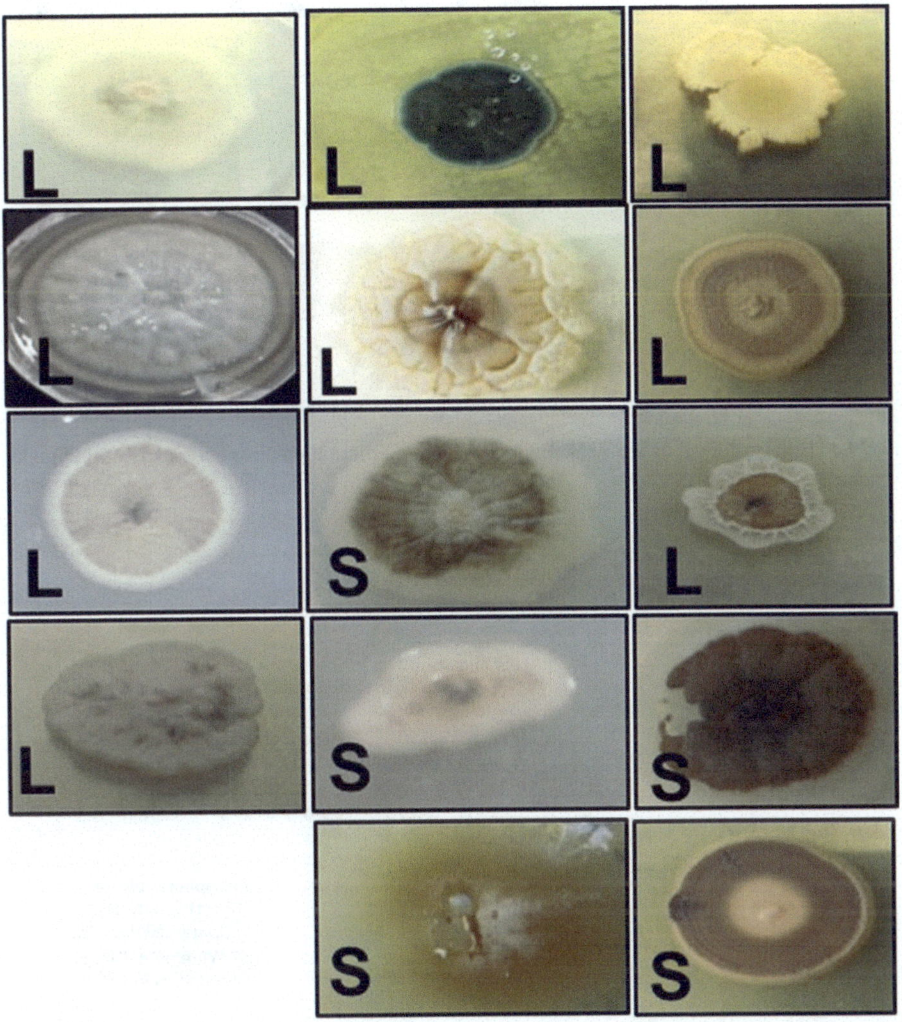

Fig. 9.3 Isolated fungal endophytes. Endophytes from different phyllospheric parts of the frankincense tree

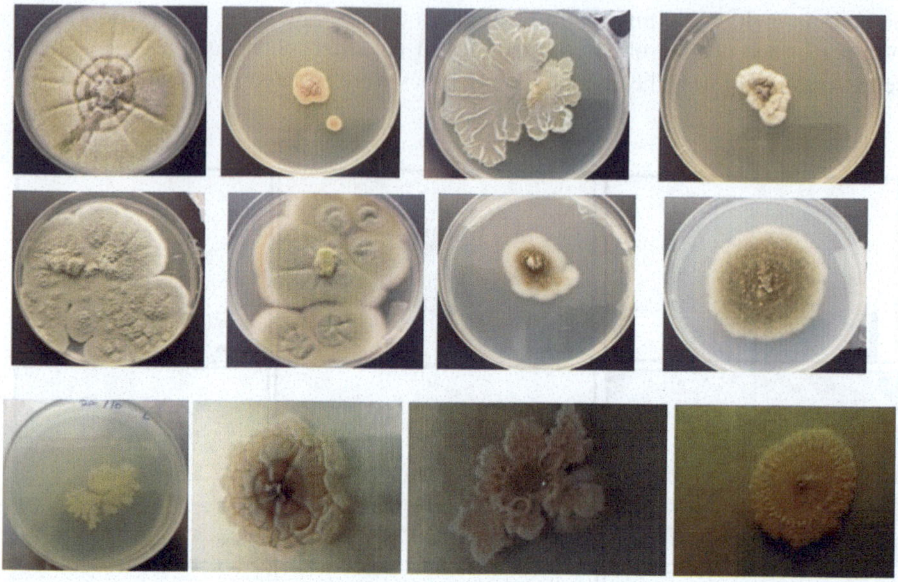

Fig. 9.4 Isolated fungal endophytes. Endophytes from different phyllospheric parts of the frankincense tree

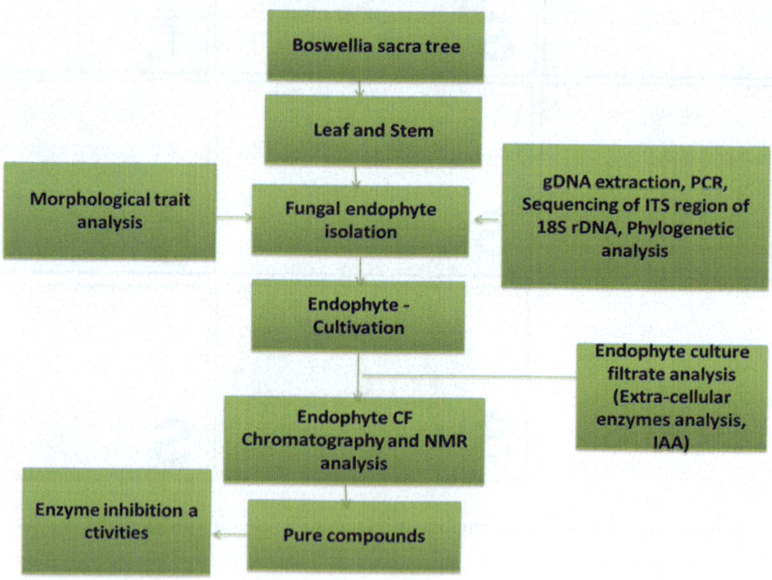

Fig. 9.5 Endophytic isolation from *Boswellia*

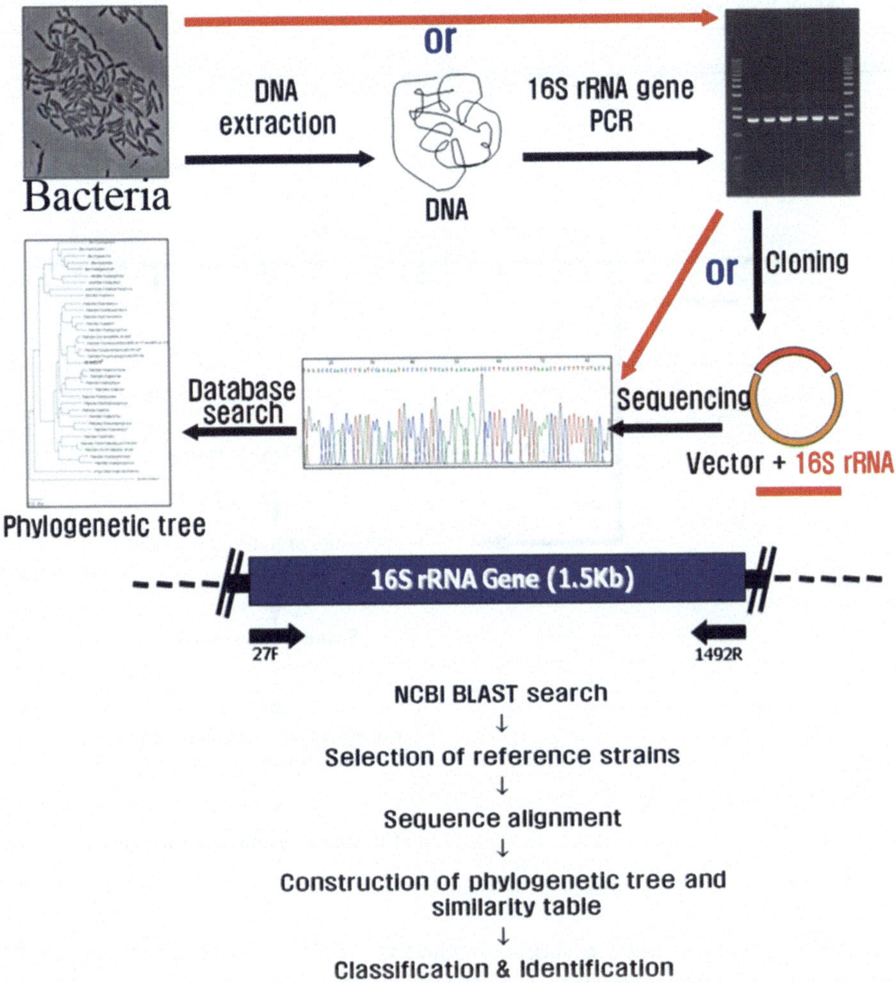

Fig. 9.6 Bacterial molecular identification methodology as adapted from Khan et al. (2012)

Molecular Identification of Endophytes Associated with *Boswellia* Trees

Most of the endophytic microbes have been identified through morphological and molecular methods. For the molecular methods, genomic DNA is extracted followed by the polymerase chain reaction (PCR) through internal transcribed spacer (ITS) or 18S rRNA region amplification (Fig. 9.6). The resulting PCR product is sequenced on a Sanger sequencer. This approach is currently a standardized method used for wide varieties of fungi isolated from diverse environments (Deshpande et al., 2016; Raja, Miller, Pearce, & Oberlies, 2017; Rosconi et al., 2013). Similarly,

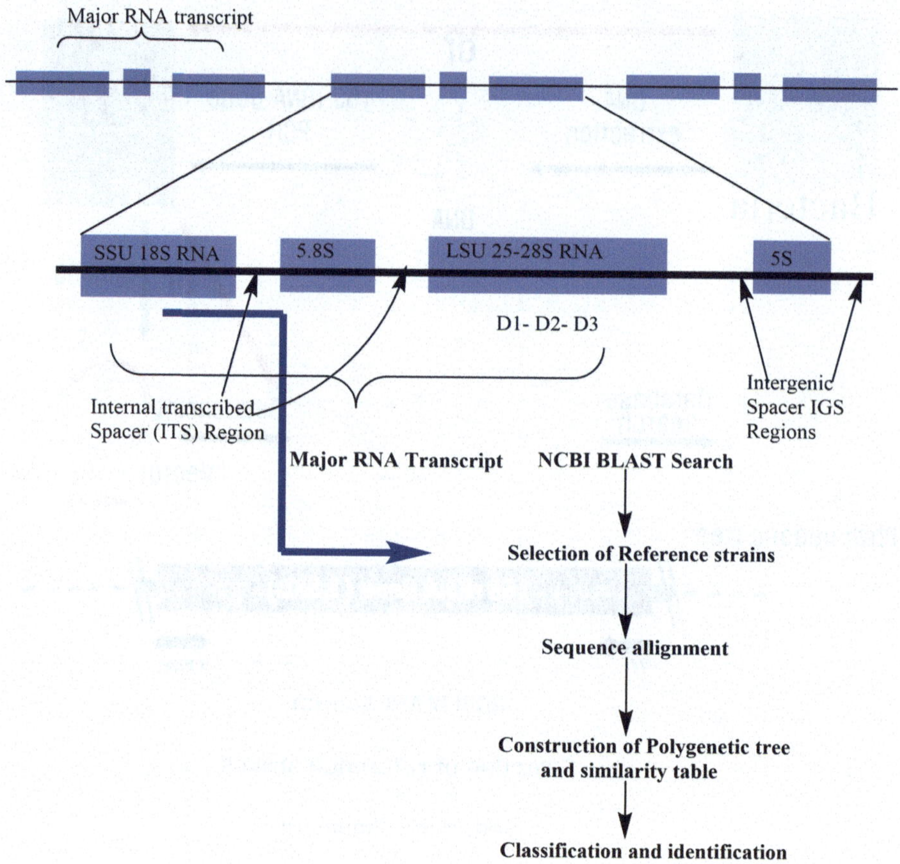

Fig. 9.7 Endophytic fungi molecular identification methodology adopted from Khan et al. (2012)

for bacterial identification, a similar method has been used. However, the PCR amplification is performed with 16S rRNA regions (Fig. 9.7; (Janda & Abbott, 2007; Patel, 2001; Sabat et al., 2017; Woo, Lau, Teng, Tse, & Yuen, 2008). The resulting sequence is cleaned from primer adopters and chimaera sequences, followed by phylogenetic analysis using MEGA 7.0 or BioEdit tools. In a phylogenetic analysis, a bootstrap value of 1000 and maximum parsimony or neighbour-joining methods have been considered ideal (Kumar & Hyde, 2004). All the sequences were aligned using pairwise CLUSTALW alignment (Tamura, Stecher, Peterson, Filipski, & Kumar, 2013). The missing or ambiguous sequences were either coded as N or deleted (Nilsson, Kristiansson, Ryberg, Hallenberg, & Larsson, 2008). The aligned sequence data were submitted to GenBank for the accession number. Similar types of methodologies have also been adopted to identify the microbes from *Boswellia*. Some of the recent studies have showed the occurrence of *Dothideomycetes* sp., *Phoma medicaginis*, *Thielavia arenaria*, *Preussia* sp. and *Aureobasidium* sp. (Table 9.1). Among these, *Preussia* sp. has

Fig. 9.8 Effects of applying endophytic fungus (*Preussia* sp. BSL10, left control; right inoculated ones) on the growth dynamics and photosynthetic pigments of *Boswellia sacra* tree saplings

been broadly distributed in the leaf and stem parts of *B. sacra* (Khan, Al-Harrasi, et al., 2016). Some of the species are common wooded fungi, e.g. *Dothideomycetes* sp., which is found extensively on tree barks (Huang et al., 2009). In addition, various strains of endophytic bacteria were also isolated and identified from *Boswellia*. These strains are *Bacillus subtilis* BSL43, *Rhizobium* sp. BSL47, *Bacillus subtilis* BSS29, *Bacillus subtilis* BSS28, *Bacillus subtilis* BSS23, *Bacillus subtilis* BSS24, *Bacillus subtilis* BSL41, *Paenibacillus popilliae* BSL27 and *Bacillus amyloliquefaciens* BSL16 (Halo et al., 2015; Khan, Al-Harrasi, et al., 2016).

Endophyte Inoculation onto *Boswellia* Trees

To understand the effects of isolated endophytes on the host *Boswellia* plants, only one study on a small proportion of trees has been performed to date. In a study by our group and reported by Khan et al. (2016), one of the isolated endophytes, *Preussia* sp. BSL10, ameliorates the leaf number, shoot length, internodes and photosynthetic apparatus (chlorophyll and carotenoids; Fig. 9.8) (Khan, Al-Harrasi, et al., 2016). These findings were also revealed by previous studies on the inoculation of endophytic fungi such as *Penicillium janthinellum* LK5, *Penicillium resedanum* LK6, *Paecilomyces formosus* LHL10, *Aureobasidium pullulan*, *Chrysosporium pseudomerdarium*, *Aspergillus fumigatus* LH02 and *P. citrinum* (Khan et al., 2012; Khan, Waqas, et al., 2015; Khan et al., 2008; Limtong et al., 2014) and endophytic bacteria such as *Paenibacillus polymyxa* (Sakayaroj, Preedanon, Supaphon, Jones, & Phongpaichit, 2010), *Bacillus subtilis* (Xu et al., 2014), *Bacillus amyloliquefaciens* (Shahzad et al., 2016), *Sphingomonas* sp. LK11 (Khan, Waqas, et al., 2015) and *Epichloë typhina* (Rozpądek et al., 2015). In addition, Ritpitakphong et al. (Khan, Shin, Jung, & Lee, 2014) suggested that applying an endophyte inoculum isolated from the leaf part can assist in improving stress tolerance against plant pathogens. That study is also connected to numerous others performed by Glick (2012), Khan et al. (2012), Khan, Hussain, et al. (2015), Khan et al. (2013), Numan et al. (2018), and White Jr and Torres (2010) on using endophytic microbes to improve plant growth and development.

Boswellia plant samples are difficult to grow because of their poor regeneration capacity and slow growth dynamics. However, numerous attempts have been made to grow it using seed germination strategies. There was a very low seed germination rate of less than 10% in most cases, which has been a barrier to plant propagation and conservation strategies. Bud growth has been suggested by Eslamieh (2010) (Eslamieh, 2010). With these difficult growth dynamics, it has been important to assess and apply plant-associated endemic microbiota, either bacteria or fungi, to improve the plant growth potential. This potential has often been dedicated to the abilities of endophytic microbes to produce bioactive secondary metabolites, phytohormones and related enzymes (Shahzad et al., 2016).

Production of Phytohormones from Endophytes Associated with *Boswellia* Trees

The production of phytohormones from endophytes can range from various types of phenolics to hormone-like compounds such as auxins (indole acetic acid, IAA) and gibberellins (Bömke & Tudzynski, 2009; Khan, Asaf, Al-Rawahi, Lee, & Al-Harrasi, 2017; Al-Hosni et al., 2018; unpublished). These phytohormones are reportedly produced by various classes of endophytic microbes (Bilal et al., 2018; Eberl, Uhe, & Unsicker, 2019; Khan, Hussain, et al., 2015). They have been considered for the green roof microbiome because of their potential to influence host growth while producing a wide array of metabolites (Fulthorpe, MacIvor, Jia, & Yasui, 2018). There are a few instances in the current literature in which endophytes have been isolated from arid-area trees.

For *Boswellia*, two recent reports by Khan et al. (2016) and Al-Hosni et al. (2018) showed that the endophytic strains *Preussia* sp. BSL10 and *Aureobasidium* sp. produce various types of phytohormones (Khan, Al-Harrasi, et al., 2016). These results were promising after the authors assessed the draft genomics (Khan, Asaf, et al., 2016) and growth-promoting functions (Khan, Al-Harrasi, et al., 2016). The results showed that both strains are producing gibberellins (GAs) and IAA. *Preussia* sp. BSL10 was producing GA_4, GA_7, GA_{15} and GA_{53}. This finding suggests the presence of non-C13 hydroxylation and C13 hydroxylation pathways in *Preussia* sp. BSL10. The GA biosynthesis could occur as follows: ent-kaurenoic acid $\rightarrow \rightarrow GA_{12} \rightarrow GA_{53} \rightarrow GA_{15} \rightarrow GA_7 \rightarrow GA_4$ (Bömke & Tudzynski, 2009). In addition, *Aureobasidium* sp. BSS6 and *Preussia* sp. BSL10 produce IAA, a class of auxin. Al-Hosni et al. (2018; unpublished) confirmed these results by showing that the gene expression and transcript accumulation pattern in BSL10 was influencing the activation of *tryptophan synthase* (a precursor for IAA synthesis), *aldehyde dehydrogenase*, indole-3-acetamide (*IAM hydrolase* encoded by *IAAH*) and *tryptophan 2-monooxygenase* (encoded by *IAAM*) followed by the further conversion of IAM to IAA (Tsavkelova et al., 2012) as has been shown in the BSL10.

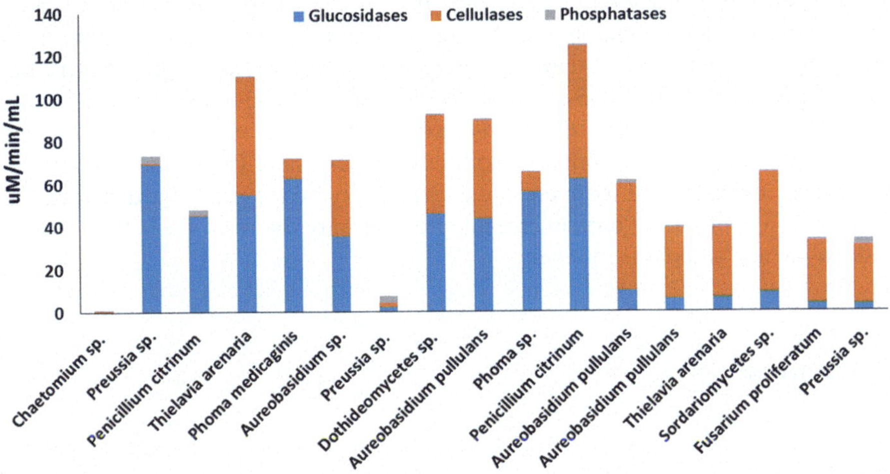

Fig. 9.9 Extracellular enzyme production ability of endophytes isolated from the different parts of frankincense trees

Extracellular Enzyme Quantification from Endophytes

Extracellular enzymes are important for converting macromolecules, for example, organic phosphate, proteins, carbohydrates and polymers, into easily transportable products (Kim et al., 2018; Wingender et al., 2012). From a mechanistic point of view, Khan, Waqas, et al. (2015) showed that exozymes can help to initiate the symbiosis process (Fig. 9.9). This type of enzyme production has been essential in preventing the breakdown of larger molecules into smaller products, but it also provides a resilient feature to combat various pathogenic infections (Esteves et al., 2014; Khan, Halo, et al., 2016; Khan et al., 2008; Petrini et al., 1993). Previous studies have revealed that endophytes isolated from arid land plants and soil are capable of producing amylases, mannanases, lipases, phenol oxidases, proteases, polyphenol oxidases, laccases, cellulases, pectin lyase and xylanases (Fouda, Hassan, Eid, & Ewais, 2015; Strobel, 2003; Sunitha, Nirmala Devi, & Srinivas, 2013; Yan et al., 2015). In the *Boswellia* species, some of the microorganism are also producing enzymes in the growth culture medium, for example, glucosidase, phosphatases and cellulases (Khan, Al-Harrasi, et al., 2016).

Future Perspectives

The genus *Boswellia* provides a great niche for exploring endophytic microbial wealth. Until the present, only the *Boswellia sacra* tree (leaves, stems and roots) has been explored for this purpose, whereas the other species have been less studied to unlock their trove of diversity. Therefore, there is a great need for further research on the isolation, identification and characterization of endophytic microflora from these species. Additionally, there are greater chances that we might identify the microbe(s) associated with resin crystallization and conversion products. Recently, the authors of this book have also tried to explore the fungal wealth in various types of resins, and they found a common *Aspergillus niger* strain. The strain adapted well to the high abundance of toxic chemical constituents in the resin. Further research is needed to explore the core microbiome of not only the resin itself but also the endophytic and ectophytic biomes.

References

Ali, S., Charles, T., & Glick, B. (2012). Delay of flower senescence by bacterial endophytes expressing 1-aminocyclopropane-1-carboxylate deaminase. *Journal of Applied Microbiology, 113*(5), 1139–1144.

Al-Hosni, K., Shahzad, R., Latif Khan, A., Muhammad Imran, Q., Al Harrasi, A., Al Rawahi, A., ... & Lee, I. J. (2018). Preussia sp. BSL-10 producing nitric oxide, gibberellins, and indole acetic acid and improving rice plant growth. *Journal of Plant Interactions, 13*(1), 112–118.

Arenal, F., Platas, G., & Pelaez, F. (2007). A new endophytic species of Preussia (Sporormiaceae) inferred from morphological observations and molecular phylogenetic analysis. *Fungal Diversity, 25*, 1–17.

Arnold, A. E., Henk, D. A., Eells, R. L., Lutzoni, F., & Vilgalys, R. (2007). Diversity and phylogenetic affinities of foliar fungal endophytes in loblolly pine inferred by culturing and environmental PCR. *Mycologia, 99*(2), 185–206.

Arnold, A. E., & Lutzoni, F. (2007). Diversity and host range of foliar fungal endophytes: Are tropical leaves biodiversity hotspots? *Ecology, 88*(3), 541–549.

Bayman, P., & Otero, J. T. (2006). Microbial endophytes of orchid roots. In *Microbial root endophytes* (pp. 153–177). Berlin, Germany: Springer.

Bilal, S., Shahzad, R., Khan, A. L., Kang, S.-M., Imran, Q. M., Al-Harrasi, A., … Lee, I.-J. (2018). Endophytic microbial consortia of phytohormones-producing fungus Paecilomyces formosus LHL10 and bacteria Sphingomonas sp. LK11 to *Glycine max* L. regulates physio-hormonal changes to attenuate aluminum and zinc stresses. *Frontiers in plant science, 9*, 1273.

Bömke, C., & Tudzynski, B. (2009). Diversity, regulation, and evolution of the gibberellin biosynthetic pathway in fungi compared to plants and bacteria. *Phytochemistry, 70*(15–16), 1876–1893.

Clay, K. (1990). Fungal endophytes of grasses. *Annual Review of Ecology and Systematics, 21*(1), 275–297.

Corrêa, R. C. G., Rhoden, S. A., Mota, T. R., Azevedo, J. L., Pamphile, J. A., de Souza, C. G. M., … Peralta, R. M. (2014). Endophytic fungi: Expanding the arsenal of industrial enzyme producers. *Journal of Industrial Microbiology & Biotechnology, 41*(10), 1467–1478.

De Bary, A. (1879). *The phenomenon of symbiosis*. Ver-lag Von Karl J. Trubner, Strasbourg, Germany.

De Battista, J., Bacon, C., Severson, R., Plattner, R., & Bouton, J. (1990). Indole acetic acid production by the fungal endophyte of tall fescue. *Agronomy Journal, 82*(5), 878–880.

Deshpande, V., Wang, Q., Greenfield, P., Charleston, M., Porras-Alfaro, A., Kuske, C. R., … Tran-Dinh, N. (2016). Fungal identification using a Bayesian classifier and the Warcup training set of internal transcribed spacer sequences. *Mycologia, 108*(1), 1–5.

Dew, R., Boissonneault, G., Gay, N., Boling, J., Cross, R., & Cohen, D. (1990). The effect of the endophyte (Acremonium coenophialum) and associated toxin (s) of tall fescue on serum titer

response to immunization and spleen cell flow cytometry analysis and response to mitogens. *Veterinary Immunology and Immunopathology, 26*(3), 285–295.
Eberl, F., Uhe, C., & Unsicker, S. B. (2019). "Friend or foe? The role of leaf-inhabiting fungal pathogens and endophytes in tree-insect interactions." *Fungal Ecology 38*, 104–112.
El-Nagerabi, S. A., Elshafie, A. E., & Alkhanjari, S. S. (2014). Endophytic fungi associated with endogenous *Boswellia sacra. Biodiversitas Journal of Biological Diversity, 15*(1), 24–30.
Eslamieh, J. (2010). Creating "Perfect" Boswellia. *Cactus and Succulent Journal, 82*(3), 126–131.
Esteves, A. C., Saraiva, M., Correia, A., & Alves, A. (2014). Botryosphaeriales fungi produce extracellular enzymes with biotechnological potential. *Canadian Journal of Microbiology, 60*(5), 332–342.
Fouda, A. H., Hassan, S. E.-D., Eid, A. M., & Ewais, E. E.-D. (2015). Biotechnological applications of fungal endophytes associated with medicinal plant Asclepias sinaica (Bioss.). *Annals of Agricultural Sciences, 60*(1), 95–104.
Fröhlich, J., & Hyde, K. D. (1999). Biodiversity of palm fungi in the tropics: Are global fungal diversity estimates realistic? *Biodiversity and Conservation, 8*(7), 977–1004.
Fulthorpe, R., MacIvor, J. S., Jia, P., & Yasui, S.-L. E. (2018). The green roof microbiome: Improving plant survival for ecosystem service delivery. *Frontiers in Ecology and Evolution, 6*, 5.
Ganley, R. J., Brunsfeld, S. J., & Newcombe, G. (2004). A community of unknown, endophytic fungi in western white pine. *Proceedings of the National Academy of Sciences, 101*(27), 10107–10112.
García, A., Rhoden, S. A., Rubin Filho, C. J., Nakamura, C. V., & Pamphile, J. A. (2012). Diversity of foliar endophytic fungi from the medicinal plant *Sapindus saponaria* L. and their localization by scanning electron microscopy. *Biological Research, 45*(2), 139–148.
Ghimire, S. R., Charlton, N. D., Bell, J. D., Krishnamurthy, Y. L., & Craven, K. D. (2011). Biodiversity of fungal endophyte communities inhabiting switchgrass (*Panicum virgatum* L.) growing in the native tallgrass prairie of northern Oklahoma. *Fungal Diversity, 47*(1), 19–27.
Glick, B. R. (2012). Plant growth-promoting bacteria: Mechanisms and applications. *Scientifica, 2012, 1*.
Göhre, V., & Robatzek, S. (2008). Breaking the barriers: Microbial effector molecules subvert plant immunity. *Annual Review of Phytopathology, 46*, 189–215.
Gunatilaka, A. L. (2006). Natural products from plant-associated microorganisms: Distribution, structural diversity, bioactivity, and implications of their occurrence. *Journal of Natural Products, 69*(3), 509–526.
Halo, B. A., Khan, A. L., Waqas, M., Al-Harrasi, A., Hussain, J., Ali, L., … Lee, I.-J. (2015). Endophytic bacteria (Sphingomonas sp. LK11) and gibberellin can improve *Solanum lycopersicum* growth and oxidative stress under salinity. *Journal of Plant Interactions, 10*(1), 117–125.
Hawksworth, D. L., & Rossman, A. Y. (1997). Where are all the undescribed fungi? *Phytopathology, 87*(9), 888–891.
Hibbett, D. S., Ohman, A., Glotzer, D., Nuhn, M., Kirk, P., & Nilsson, R. H. (2011). Progress in molecular and morphological taxon discovery in Fungi and options for formal classification of environmental sequences. *Fungal Biology Reviews, 25*(1), 38–47.
Higgins, K. L., Arnold, A. E., Miadlikowska, J., Sarvate, S. D., & Lutzoni, F. (2007). Phylogenetic relationships, host affinity, and geographic structure of boreal and arctic endophytes from three major plant lineages. *Molecular Phylogenetics and Evolution, 42*(2), 543–555.
Huang, W., Cai, Y., Surveswaran, S., Hyde, K., Corke, H., & Sun, M. (2009). Molecular phylogenetic identification of endophytic fungi isolated from three Artemisia species. *Fungal diversity, 36*, 69–88.
Janda, J. M., & Abbott, S. L. (2007). 16S rRNA gene sequencing for bacterial identification in the diagnostic laboratory: Pluses, perils, and pitfalls. *Journal of Clinical Microbiology, 45*(9), 2761–2764.
Khan, A. L., Al-Harrasi, A., Al-Rawahi, A., Al-Farsi, Z., Al-Mamari, A., Waqas, M., … Shin, J.-H. (2016). Endophytic fungi from Frankincense tree improves host growth and produces extracellular enzymes and indole acetic acid. *PLoS One, 11*(6), e0158207.

Khan, A. L., Al-Harrasi, A., Shahzad, R., Imran, Q. M., Yun, B.-W., Kim, Y.-H., … Lee, I.-J. (2018). Regulation of endogenous phytohormones and essential metabolites in frankincense-producing *Boswellia sacra* under wounding stress. *Acta Physiologiae Plantarum, 40*(6), 113.

Khan, A. L., Asaf, S., Al-Rawahi, A., Lee, I.-J., & Al-Harrasi, A. (2017). Rhizospheric microbial communities associated with wild and cultivated frankincense producing *Boswellia sacra* tree. *PLoS One, 12*(10), e0186939.

Khan, A. L., Asaf, S., Khan, A. R., Al-Harrasi, A., Al-Rawahi, A., & Lee, I.-J. (2016). First draft genome sequencing of indole acetic acid producing and plant growth promoting fungus Preussia sp. BSL10. *Journal of Biotechnology, 225*, 44–45.

Khan, A. L., Halo, B. A., Elyassi, A., Ali, S., Al-Hosni, K., Hussain, J., … Lee, I.-J. (2016). Indole acetic acid and ACC deaminase from endophytic bacteria improves the growth of *Solanum lycopersicum*. *Electronic Journal of Biotechnology, 21*, 58–64.

Khan, A. L., Hamayun, M., Kang, S.-M., Kim, Y.-H., Jung, H.-Y., Lee, J.-H., & Lee, I.-J. (2012). Endophytic fungal association via gibberellins and indole acetic acid can improve plant growth under abiotic stress: An example of Paecilomyces formosus LHL10. *BMC Microbiology, 12*(1), 3.

Khan, A. L., Hussain, J., Al-Harrasi, A., Al-Rawahi, A., & Lee, I.-J. (2015). Endophytic fungi: Resource for gibberellins and crop abiotic stress resistance. *Critical Reviews in Biotechnology, 35*(1), 62–74.

Khan, A. L., & Lee, I.-J. (2013). Endophytic Penicillium funiculosum LHL06 secretes gibberellin that reprograms *Glycine max* L. growth during copper stress. *BMC Plant Biol, 13*(1), 86.

Khan, A. L., Shin, J.-H., Jung, H.-Y., & Lee, I.-J. (2014). Regulations of capsaicin synthesis in *Capsicum annuum* L. by Penicillium resedanum LK6 during drought conditions. *Scientia Horticulturae, 175*, 167–173.

Khan, A. L., Waqas, M., Asaf, S., Kamran, M., Shahzad, R., Bilal, S., … Yun, B.-W. (2017). Plant growth-promoting endophyte Sphingomonas sp. LK11 alleviates salinity stress in *Solanum pimpinellifolium*. *Environmental and Experimental Botany, 133*, 58–69.

Khan, A. L., Waqas, M., Hamayun, M., Al-Harrasi, A., Al-Rawahi, A., & Lee, I.-J. (2013). Co-synergism of endophyte Penicillium resedanum LK6 with salicylic acid helped *Capsicum annuum* in biomass recovery and osmotic stress mitigation. *BMC Microbiology, 13*(1), 51.

Khan, A. L., Waqas, M., Hussain, J., Al-Harrasi, A., Hamayun, M., & Lee, I.-J. (2015). Phytohormones enabled endophytic fungal symbiosis improve aluminum phytoextraction in tolerant *Solanum lycopersicum*: An examples of Penicillium janthinellum LK5 and comparison with exogenous GA3. *Journal of Hazardous Materials, 295*, 70–78.

Khan, S. A., Hamayun, M., Yoon, H., Kim, H.-Y., Suh, S.-J., Hwang, S.-K., … Yoon, U.-H. (2008). Plant growth promotion and Penicillium citrinum. *BMC Microbiology, 8*(1), 231.

Kharwar, R. N., Mishra, A., Gond, S. K., Stierle, A., & Stierle, D. (2011). Anticancer compounds derived from fungal endophytes: Their importance and future challenges. *Natural Product Reports, 28*(7), 1208–1228.

Kim, Y., Seo, C. W., Khan, A. L., Mun, B. G., Shahzad, R., Ko, J. W., ... & Lee, I. J. (2018). Exo-ethylene application mitigates waterlogging stress in soybean (Glycine max L.). *BMC plant biology, 18*(1), 254–270.

Krings, M., Taylor, T. N., Hass, H., Kerp, H., Dotzler, N., & Hermsen, E. J. (2007). Fungal endophytes in a 400-million-yr-old land plant: Infection pathways, spatial distribution, and host responses. *New Phytologist, 174*(3), 648–657.

Kuldau, G., & Bacon, C. (2008). Clavicipitaceous endophytes: Their ability to enhance resistance of grasses to multiple stresses. *Biological Control, 46*(1), 57–71.

Kumar, D. S. S., & Hyde, K. D. (2004). Biodiversity and tissue-recurrence of endophytic fungi in Tripterygium wilfordii. *Fungal diversity, 17*, 69–90.

Kusari, S., Hertweck, C., & Spiteller, M. (2012). Chemical ecology of endophytic fungi: Origins of secondary metabolites. *Chemistry & Biology, 19*(7), 792–798.

Kusari & Spiteller (2012). Metabolomics of endophytic fungi producing associated plant secondary metabolites: progress, challenges and opportunities U. Roessner (Ed.), Metabolomics, InTech, Rijeka, Croatia, pp. 241–266.

Kusari, S., Zühlke, S., & Spiteller, M. (2009). An endophytic fungus from *Camptotheca acuminata* that produces camptothecin and analogues. *Journal of Natural Products, 72*(1), 2–7.

Lewis, D. H. (1985). *Symbiosis and mutualism: crisp concepts and soggy semantics*. In: Boucher, D. H., (ed). The biology of mutualism. London, UK: Croom Helm Ltd, 29–39.

Limtong, S., Kaewwichian, R., Yongmanitchai, W., & Kawasaki, H. (2014). Diversity of culturable yeasts in phylloplane of sugarcane in Thailand and their capability to produce indole-3-acetic acid. *World Journal of Microbiology and Biotechnology, 30*(6), 1785–1796.

Lindow, S. E., & Brandl, M. T. (2003). Microbiology of the phyllosphere. *Applied and Environmental Microbiology, 69*(4), 1875–1883.

Liu, X., Dong, M., Chen, X., Jiang, M., Lv, X., & Yan, G. (2007). Antioxidant activity and phenolics of an endophytic Xylaria sp. from *Ginkgo biloba*. *Food Chemistry, 105*(2), 548–554.

Lodge, D. J., Fisher, P., & Sutton, B. (1996). Endophytic fungi of *Manilkara bidentata* leaves in Puerto Rico. *Mycologia, 88*, 733–738.

Maheshwari, R. (2006). What is an endophytic fungus. *Current Science, 90*(10), 1309.

Müller, C. B., & Krauss, J. (2005). Symbiosis between grasses and asexual fungal endophytes. *Current Opinion in Plant Biology, 8*(4), 450–456.

Nilsson, R. H., Kristiansson, E., Ryberg, M., Hallenberg, N., & Larsson, K.-H. (2008). Intraspecific ITS variability in the kingdom Fungi as expressed in the international sequence databases and ITS implications for molecular species identification. *Evolutionary Bioinformatics Online, 4*, S653.

Numan, M., Bashir, S., Khan, Y., Mumtaz, R., Shinwari, Z. K., Khan, A. L., … Ahmed, A.-H. (2018). Plant growth promoting Bacteria as an alternative strategy for salt tolerance in plants: A review. *Microbiological Research, 209*, 21.

Patel, J. B. (2001). 16S rRNA gene sequencing for bacterial pathogen identification in the clinical laboratory. *Molecular Diagnosis, 6*(4), 313–321.

Petrini, O., Sieber, T. N., Toti, L., & Viret, O. (1993). Ecology, metabolite production, and substrate utilization in endophytic fungi. *Natural Toxins, 1*(3), 185–196.

Pirozynski, K., & Malloch, D. (1975). The origin of land plants: A matter of mycotrophism. *Biosystems, 6*(3), 153–164.

Porras-Alfaro, A., Herrera, J., Sinsabaugh, R. L., Odenbach, K. J., Lowrey, T., & Natvig, D. O. (2008). Novel root fungal consortium associated with a dominant desert grass. *Applied and Environmental Microbiology, 74*(9), 2805–2813.

Raja, H. A., Miller, A. N., Pearce, C. J., & Oberlies, N. H. (2017). Fungal identification using molecular tools: A primer for the natural products research community. *Journal of Natural Products, 80*(3), 756–770.

Redecker, D., Kodner, R., & Graham, L. E. (2000). Glomalean fungi from the Ordovician. *Science, 289*(5486), 1920–1921.

Robert-Seilaniantz, A., Navarro, L., Bari, R., & Jones, J. D. (2007). Pathological hormone imbalances. *Current Opinion in Plant Biology, 10*(4), 372–379.

Rodriguez, R. J., Henson, J., Van Volkenburgh, E., Hoy, M., Wright, L., Beckwith, F., … Redman, R. S. (2008). Stress tolerance in plants via habitat-adapted symbiosis. *The ISME Journal, 2*(4), 404.

Rosconi, F., Davyt, D., Martínez, V., Martínez, M., Abin-Carriquiry, J. A., Zane, H., … Fabiano, E. (2013). Identification and structural characterization of serobactins, a suite of lipopeptide siderophores produced by the grass endophyte H erbaspirillum seropedicae. *Environmental Microbiology, 15*(3), 916–927.

Rowan, D. D., Latch, G. C., Bacon, C., & White, J. (1994). Utilization of endophyte-infected perennial ryegrasses for increased insect resistance. In *Biotechnology of endophytic fungi of grasses* (pp. 169–183). Boca Raton, London/New York: CRC Press.

Rozpądek, P., Wężowicz, K., Nosek, M., Ważny, R., Tokarz, K., Lembicz, M., … Turnau, K. (2015). The fungal endophyte Epichloë typhina improves photosynthesis efficiency of its host orchard grass (*Dactylis glomerata*). *Planta, 242*(4), 1025–1035.

Sabat, A. J., Zanten, E., Akkerboom, V., Wisselink, G., Slochteren, K., Boer, R. F., … Kooistra-Smid, A. M. M. (2017). Targeted next-generation sequencing of the 16S-23S rRNA region for culture-independent bacterial identification-increased discrimination of closely related species. *Scientific Reports, 7*(1), 3434.

Saikkonen, K., Wäli, P., Helander, M., & Faeth, S. H. (2004). Evolution of endophyte–plant symbioses. *Trends in Plant Science, 9*(6), 275–280.

Sakayaroj, J., Preedanon, S., Supaphon, O., Jones, E. G., & Phongpaichit, S. (2010). Phylogenetic diversity of endophyte assemblages associated with the tropical seagrass *Enhalus acoroides* in Thailand. *Fungal Diversity, 42*(1), 27–45.

Santamaría, J., & Bayman, P. (2005). Fungal epiphytes and endophytes of coffee leaves (*Coffea arabica*). *Microbial Ecology, 50*(1), 1–8.

Sapp, J. (2004). The dynamics of symbiosis: An historical overview. *Canadian Journal of Botany, 82*(8), 1046–1056.

Schulz, B., & Boyle, C. (2005). The endophytic continuum. *Mycological Research, 109*(6), 661–686.

Sessitsch, A., Hardoim, P., Döring, J., Weilharter, A., Krause, A., Woyke, T., … Rahalkar, M. (2012). Functional characteristics of an endophyte community colonizing rice roots as revealed by metagenomic analysis. *Molecular Plant-Microbe Interactions, 25*(1), 28–36.

Shahzad, R., Waqas, M., Khan, A. L., Asaf, S., Khan, M. A., Kang, S.-M., … Lee, I.-J. (2016). Seed-borne endophytic *Bacillus amyloliquefaciens* RWL-1 produces gibberellins and regulates endogenous phytohormones of *Oryza sativa*. *Plant Physiology and Biochemistry, 106*, 236–243.

Soussi, A., Ferjani, R., Marasco, R., Guesmi, A., Cherif, H., Rolli, E., … Cherif, A. (2016). Plant-associated microbiomes in arid lands: Diversity, ecology and biotechnological potential. *Plant and Soil, 405*(1–2), 357–370.

Stierle, A., Strobel, G., & Stierle, D. (1993). Taxol and taxane production by Taxomyces andreanae, an endophytic fungus of Pacific yew. *Science, 260*(5105), 214–216.

Stone, J. K., Bacon, C. W., & White, J. F. (2000). An overview of endophytic mibrobes: Endophytism defined. In: Microbial Endophytes. Bacon, C. W., & White, J. F. (eds.). Marcel Dekker, New York, p. 3–30.

Strobel, G. A. (2003). Endophytes as sources of bioactive products. *Microbes and Infection, 5*(6), 535–544.

Sun, P.-F., Fang, W.-T., Shin, L.-Y., Wei, J.-Y., Fu, S.-F., & Chou, J.-Y. (2014). Indole-3-acetic acid-producing yeasts in the phyllosphere of the carnivorous plant Drosera indica L. *PLoS One, 9*(12), e114196.

Sun, Y., Wang, Q., Lu, X., Okane, I., & Kakishima, M. (2011). Endophytic fungi associated with two Suaeda species growing in alkaline soil in China. *Mycosphere, 2*(3), 239–248.

Sunitha, V., Nirmala Devi, D., & Srinivas, C. (2013). Extracellular enzymatic activity of endophytic fungal strains isolated from medicinal plants. *World Journal of Agricultural Sciences, 9*(1), 1–9.

Tamura, K., Stecher, G., Peterson, D., Filipski, A., & Kumar, S. (2013). MEGA6: Molecular evolutionary genetics analysis version 6.0. *Molecular Biology and Evolution, 30*(12), 2725–2729.

Tolera, M., Sass-Klaassen, U., Eshete, A., Bongers, F., & Sterck, F. J. (2013). Frankincense tree recruitment failed over the past half century. *Forest Ecology and Management, 304*, 65–72.

Tsavkelova, E., Oeser, B., Oren-Young, L., Israeli, M., Sasson, Y., Tudzynski, B., & Sharon, A. (2012). Identification and functional characterization of indole-3-acetamide-mediated IAA biosynthesis in plant-associated Fusarium species. *Fungal Genetics and Biology, 49*(1), 48–57.

Van Der Heijden, M. G., Bardgett, R. D., & Van Straalen, N. M. (2008). The unseen majority: Soil microbes as drivers of plant diversity and productivity in terrestrial ecosystems. *Ecology Letters, 11*(3), 296–310.

Voříšková, J., & Baldrian, P. (2013). Fungal community on decomposing leaf litter undergoes rapid successional changes. *The ISME Journal, 7*(3), 477.

Wennstrom, A. (1994). Endophyte: the misuse of an old term. *Oikos, 71*, 535–536.

White, J. F., Jr., & Torres, M. S. (2010). Is plant endophyte-mediated defensive mutualism the result of oxidative stress protection? *Physiologia Plantarum, 138*(4), 440–446.

Wilson, T. M. (1993). Strategies to protect crop plants against viruses: Pathogen-derived resistance blossoms. *Proceedings of the National Academy of Sciences, 90*(8), 3134–3141.

Wingender, G., Stepniak, D., Krebs, P., Lin, L., McBride, S., Wei, B., … Kronenberg, M. (2012). Intestinal microbes affect phenotypes and functions of invariant natural killer T cells in mice. *Gastroenterology, 143*(2), 418–428.

Woo, P., Lau, S., Teng, J., Tse, H., & Yuen, K.-Y. (2008). Then and now: Use of 16S rDNA gene sequencing for bacterial identification and discovery of novel bacteria in clinical microbiology laboratories. *Clinical Microbiology and Infection, 14*(10), 908–934.

Xu, M., Sheng, J., Chen, L., Men, Y., Gan, L., Guo, S., & Shen, L. (2014). Bacterial community compositions of tomato (Lycopersicum esculentum Mill.) seeds and plant growth promoting activity of ACC deaminase producing Bacillus subtilis (HYT-12-1) on tomato seedlings. *World Journal of Microbiology and Biotechnology, 30*(3), 835–845.

Yan, J., Broughton, S., Yang, S., & Gange, A. (2015). Do endophytic fungi grow through their hosts systemically? *Fungal Ecology, 13*, 53–59.

Chapter 10
Resin Composition and Structural Diversity

Introduction

The twentieth century was witness to a remarkable interest not only from chemists but also from biologists from different specialties who were trying to understand how living organisms produce secondary metabolites. This interest was also boosted by the discovery of the antibiotics that conferred trust in natural products that in turn played a pivotal role in emerging pharmaceutical industries. In that century, it became possible to isolate active ingredients from plants for the first time, although plants have been used for therapeutic purposes throughout the history of mankind.

Terpenes or terpenoids form the major class of secondary metabolites that are biosynthesized by plants. More than 20,000 triterpenes were reported (Hill & Connolly, 2012), which makes them the largest group of natural products to be isolated from the plant kingdom, a fact that supports the enormous structural diversity of this important class of compounds. The 19 known species of *Boswellia* have contributed to the biologically active natural products isolated from the plant kingdom due to the rich contents of terpenes in their resins.

Natural products have been isolated from their natural sources for different applications, including but not limited to flavour and fragrance (e.g. citral from lemongrass oil, *Cymbopogon citratus*, or linalool from lavender oil, *Lavandula angustifolia*), pharmaceutical drugs (e.g. Taxol®, which was isolated from *Taxus brevifolia*; morphine, isolated from opium, *Papaver somniferum* and *P. setigerum*; quinine, isolated from the bark of the cinchona tree; and salicin from the bark of the willow tree) and agrochemicals (insecticides) (e.g. pyrethrins from *Chrysanthemum cinerariifolium*). The *Boswellia* genus is considered very promising due to its rich boswellic acid contents, which are known for their anti-inflammatory activities (Al-Harrasi, Hussain, Csuk, & Khan, 2018).

In general, with variations in different *Boswellia* species, the contents of the resins isolated from different species and different geographical locations of the same species consist of volatile components and acidic fractions in addition to

A. Al-Harrasi et al., *Biology of Genus Boswellia*,
https://doi.org/10.1007/978-3-030-16725-7_10

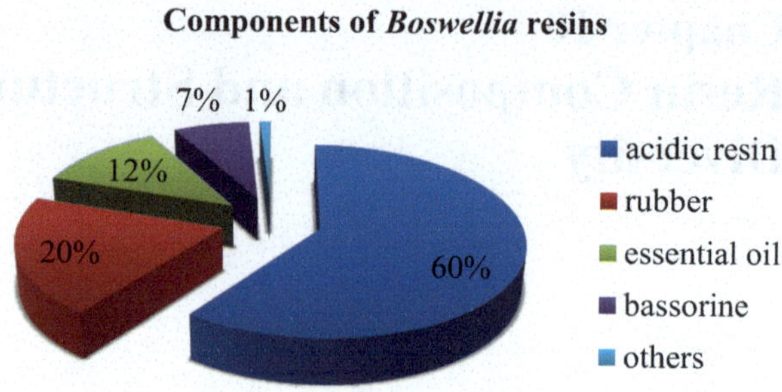

Fig. 10.1 Different components of *Boswellia* resin

Table 10.1 Naturally isolated and synthetically prepared diterpenes and triterpenes from *Boswellia* resins

Terpenes in the *Boswellia* genus	Number
Cembranes and verticillanes	34
Ursanes and oleananes	27
Lupanes	26
Boswellic acids and their derivatives	18
Dammaranes	11
Tirucallic acids	9
Synthetic boswellic acids	215

other components (Fig. 10.1). The acidic fraction has attracted the attention of medicinal chemists more than other components due to the known biological activities of the resin.

In the essential oil of *Boswellia*, more than 300 components have been identified, and the resin contains diterpenes (34) and triterpenes (91) (Al-Harrasi, Hussain, et al., 2018). This number shows the structurally complex diversity of this important genus (Table 10.1). Synthetic chemists have contributed a great deal to further diversify the naturally isolated terpenes by chemical modification through partial synthesis to enhance their biological activities, and there are more than 200 reported synthetic derivatives, mostly from boswellic acids (Al-Harrasi, Hussain, et al., 2018).

The essential oils obtained from the resins of different *Boswellia* species as reported by many authors vary from one species to another (Mertens, Buettner, & Kirchhoff, 2009). This variation depends on many factors including the botanical origin, geographical location, time of harvest and other environmental factors. This variation in the essential oil composition coupled with the rich terpene content has enriched the structural diversity of this historical resin.

This chapter highlights the structural diversity of the *Boswellia* genus. The chemistry, including the isolation and synthesis as well as the biological activities of boswellic acids and other terpenoids, is not discussed in this chapter and can be found on the book *Chemistry and Bioactivity of Boswellic acids and Other Terpenoids of the Genus Boswellia* (Al-Harrasi, Hussain, et al., 2018).

Structural Diversity

Following the development of the isoprene rule by Wallach (Nobel Prize in Chemistry 1910) and Ruzicka (Nobel Prize in Chemistry 1939), the road was paved for the elucidation of many natural terpenes. A terpene comprises C5 units (isoprene units) that make up the hydrocarbon skeleton of terpenes. Figure 10.2 represents the basic hydrocarbons of terpenes with given examples from the genus *Boswellia*. In this genus, monoterpenes (C10), sesquiterpenes (C15), diterpenes (C20) and triterpenes (C30) were isolated.

To define the basic structures of the terpenes, the isopropyl part of 2-methylbutane is defined as the head, whereas the ethyl part is defined as the tail. For monoterpenes, sesquiterpenes and diterpenes, the units are connected head-to-tail, whereas the triterpenes are connected from tail-to-tail (Breitmaier, 2006).

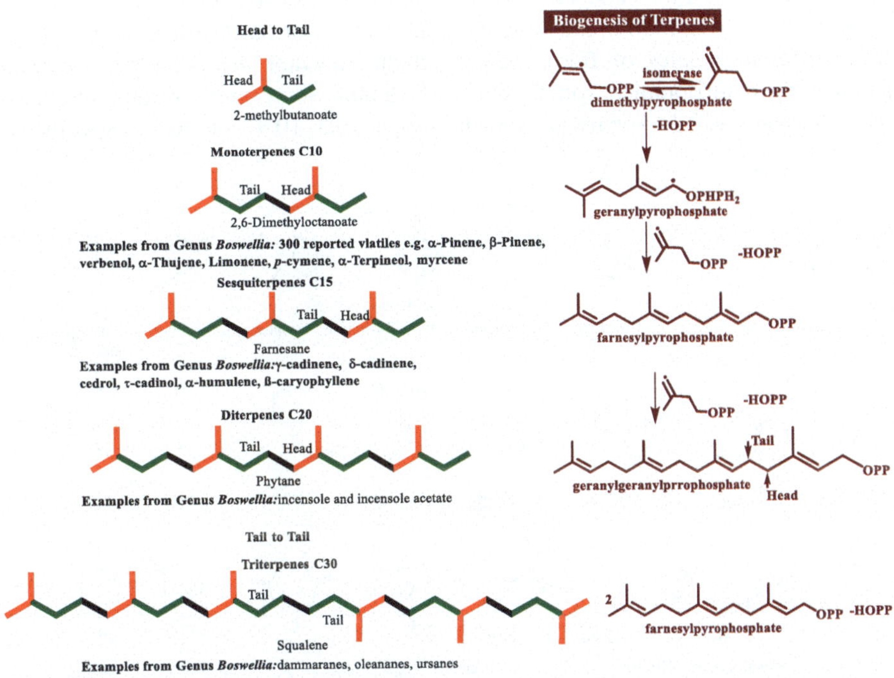

Fig. 10.2 Basic hydrocarbons of the terpenes with examples from the *Boswellia* genus

Structural Diversity in *Boswellia* Essential Oil

Essential oils are mixtures of volatile compounds that are excreted by glandular hairs or deposited in the plant body in cell organelles, idioblasts (oil cells or bodies), schizogenous or lysigenous cavities or ducts (Worwood, 2016). These oils are found in various parts of the plant in varying amounts and are obtained by hydrodistillation. They comprise monoterpenes, sesquiterpenes, a few diterpenes and some aromatic compounds. Frankincense essential oil from most of the known *Boswellia* species has been analysed extensively (Mertens et al., 2009), given its fragrant and therapeutic importance.

In the *Boswellia* genus, the production of the resin occurs at the cambium (total boswellic acids 1.0%) and epidermis (total boswellic acids 4.93%) due to the existence of the resin-producing canals. The micrographs clearly demarcate the presence of radial and axial resin canals in the inner bark region (between the epidermis and vascular cambium) of *Boswellia sacra* (Al-Harrasi, Rehman, et al., 2018) and the presence of resinous ducts in *Boswellia papyrifera* (Tolera et al., 2012).

More than 300 volatile constituents have been reported (Mertens et al., 2009) from different species of frankincense, which indicates the complex structural nature of the resin obtained from the *Boswellia* genus. Some common monoterpenes, sesquiterpenes and diterpenes are shown in Figs. 10.3, 10.4, and 10.5, respectively.

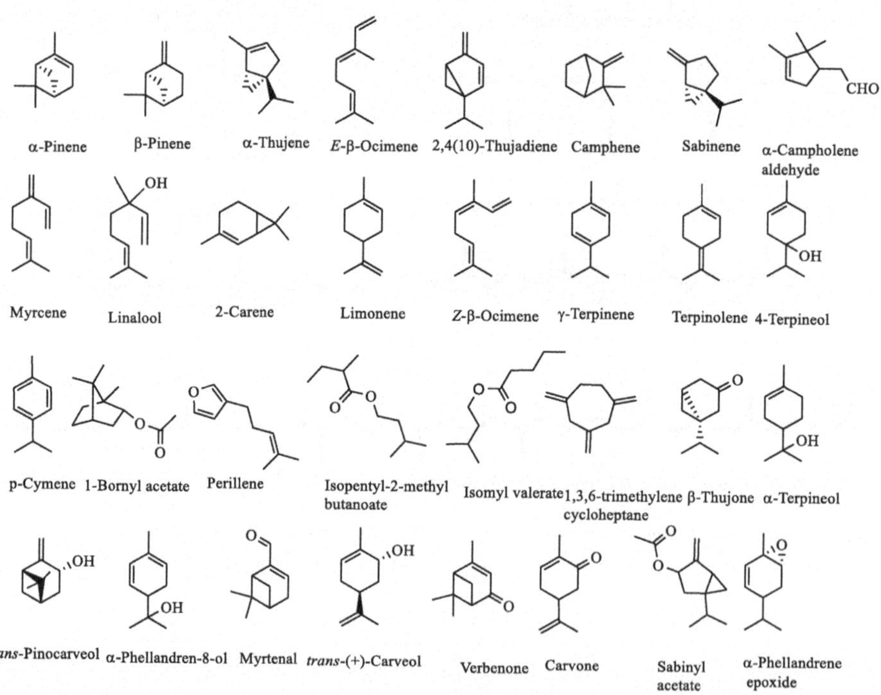

Fig. 10.3 Common monoterpenes detected in different *Boswellia* species

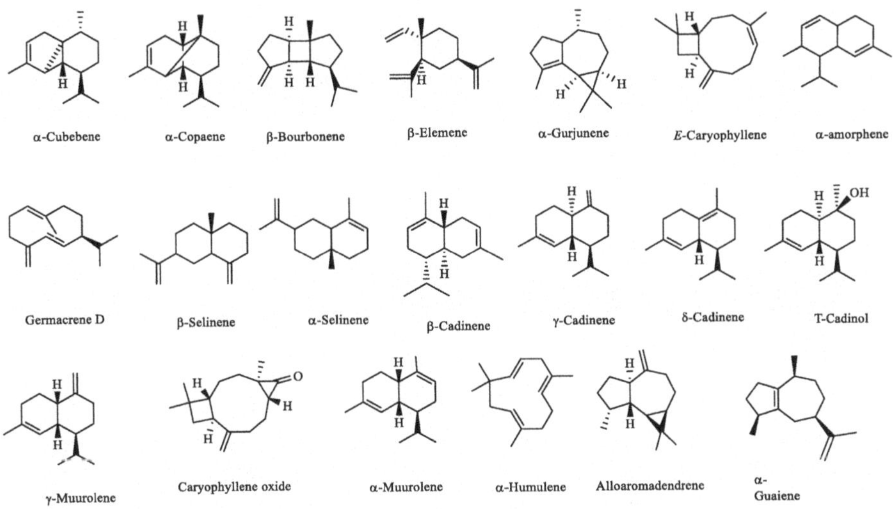

Fig. 10.4 Common sesquiterpenes detected in different *Boswellia* species

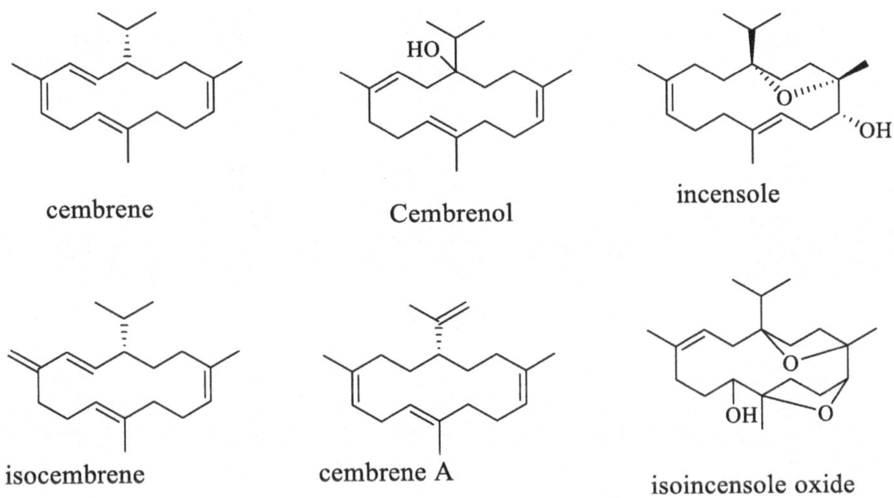

Fig. 10.5 Common diterpenes detected in different *Boswellia* species

It is challenging to find specific markers for *Boswellia* species by using the chemical profile of the essential oil due to the substantial overlap of the constituents despite the variation in the percentages of each constituent. Despite this difficulty, there are a few terpenes that could be identified as predominant in each studied species that are summarized in Fig. 10.6.

In some literature, chemotaxonomically and botanically, *Boswellia carterii* is often regarded as a synonym of *Boswellia sacra* (Hamm, Bleton, Connan, & Tchapla, 2005; Thulin & Warfa, 1987). For the essential oil analysis, the two species

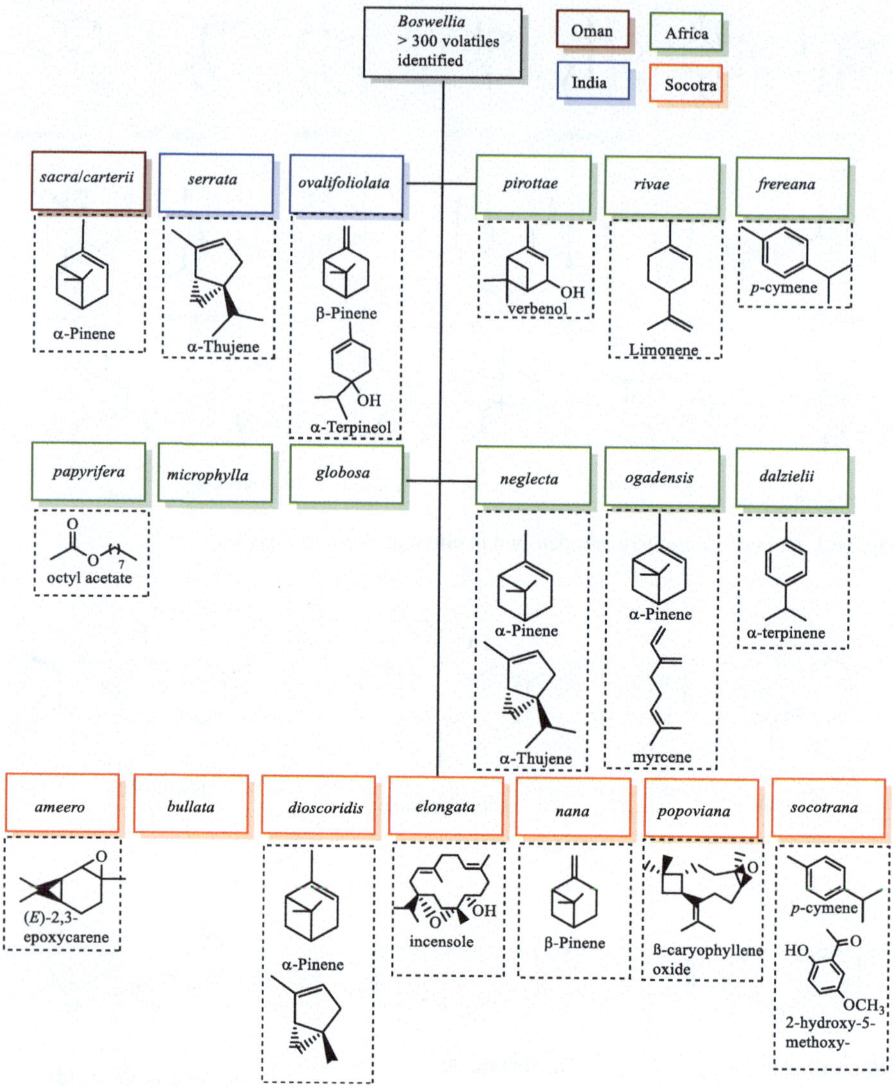

Fig. 10.6 Predominant compounds in essential oils from different *Boswellia* species

are known as synonyms in which α-pinene (Camarda, Dayton, Di Stefano, Pitonzo, & Schillaci, 2007; Murthy, Reddy, Rani, & Pullaiah, 2016) is predominant despite the difference in its bioavailability in *Boswellia sacra* and *Boswellia carterii*.

Location-wise, India hosts two species of frankincense, namely, *Boswellia serrata* and *Boswellia ovalifoliolata*, which shares *Boswellia serrata* with Pakistan and Sri Lanka and *Boswellia ovalifoliolata* with Sri Lanka (Small, 2017). *Boswellia serrata* was the most-studied species in the *Boswellia* genus, and several monoterpenes were found to be shared with all the analysed samples. α-Thujene (Singh, Kumar, Bhandari, Pathania, & Lal, 2007) was the most abundant in addition to β-bourbonene, 3-carene, *p*-cymene, limonene, linalool, α-phellandrene, α-pinene, sabinene and γ-terpinene (Mertens et al., 2009); β-pinene and α-terpineol are the most abundant monoterpenes in the essential oil of *Boswellia ovalifoliolata* (Fig. 10.6).

In Africa, there are nine known species of *Boswellia* (Fig. 10.6). The essential oil of *Boswellia pirottae* from Ethiopia [9] is found to have *trans*-verbenol as the primary monoterpene (Başer, Demirci, Dekebo, & Dagne, 2003). *Boswellia rivae* grows in Kenya, Somalia and Ethiopia (Eslamieh, 2011), and limonene is its predominant monoterpene (Başer et al., 2003). It is interesting to note that limonene is found to be shared among all the studied *Boswellia* species (Mertens et al., 2009). Camphene, limonene, α-pinene and α-terpineol are common monoterpenes in the three African samples, primarily *Boswellia papyrifera*, *Boswellia pirottae* and *Boswellia frereana* (Başer et al., 2003; Strappaghetti, Corsano, Craveiro, & Proietti, 1982). Limonene contributes to the overall fragrance of frankincense oil due to its presence in almost all the analysed frankincense species (Mertens et al., 2009). The essential oil of *Boswellia neglecta* is characterized by the predominance of α-pinene and α-thujene (Başer et al., 2003; Provan, Gray, & Waterman, 1987). The hydrodistillate of *Boswellia dalzielii* Hutch. leaves presents α-pinene (45.7%) and α-terpinene (11.5%) as the predominant compounds. The abundance of monoterpenoids in this oil is characteristic of this genus (Kubmarawa, Ogunwande, Okorie, Olawore, & Kasali, 2006).

Socotra in Yemen hosts seven rare species of *Boswellia*. These species are *Boswellia popoviana*, *Boswellia nana*, *Boswellia bullata*, *Boswellia dioscorides*, *Boswellia ameero*, *Boswellia elongata* and *Boswellia socotrana*. Their volatile constituents vary from one to another. *Boswellia ameero* is characterized by epoxy-carene (Ali et al., 2008). *Boswellia dioscorides* and *Boswellia socotrana* are characterized by high contents of monoterpenes (Mothana, Hasson, Schultze, Mowitz, & Lindequist, 2011), whereas *Boswellia elongata* contains a high percentage of diterpenes in its particular incensole (Ali et al., 2008). The essential oil of *Boswellia nana* possesses a high concentration of α-pinene (Adams, 1997). Caryophyllene oxide constitutes the largest percentage of the essential oil from *Boswellia popoviana* (Adams, 1997).

Niebler et al. (Niebler & Buettner, 2016; Niebler, Eslamieh, & Buettner, 2016) have thoroughly investigated the essential oils of *Boswellia*, and so far, their work can be regarded as the most reliable reference given the reported scattered data and the known controversy in *Boswellia* species. α-Pinene and limonene were found to be the only common compounds present in all the samples from all the investigated *Boswellia* varieties, as analysed by means of hydrodistillation (Niebler et al., 2016).

Structural Diversity in *Boswellia* Resins

Some species in the *Boswellia* genus are known to be rich in diterpenes, particularly, cembranes. Several cembranes resulting from the cyclization of geranylgeranyl pyrophosphate (GGPP) were reported in the genus *Boswellia* (Al-Harrasi, Hussain, et al., 2018). Furthermore, a few members of the verticillane-type diterpenes were reported in *Boswellia carterii* (Basar, 2005; Paul, 2012), and they can be considered as precursors of the taxane group of diterpenes. They originate from geranylgeranyl diphosphate and are derived from cembrane derivatives via 11,15-cyclization. Due to taxonomic controversy, the natural source of some of these cembranes was in fact *Boswellia papyrifera* and not *Boswellia carterii*, as in the case of incensole and incensole acetate (Al-Harrasi, Hussain, et al., 2018).

Triterpenes are usually synthesized in plant tissues during developmental stages. Their production is also influenced in response to exposures to abiotic and/or biotic stresses or due to elicitors such as methyl jasmonate. The cyclization of the historical precursor 2,3-oxidosqualene led to more than 100 triterpenic scaffolds that were biosynthesized by plants, which resulted in more than 20,000 reported triterpenes to date. These building blocks served as precursors for further partial syntheses by the

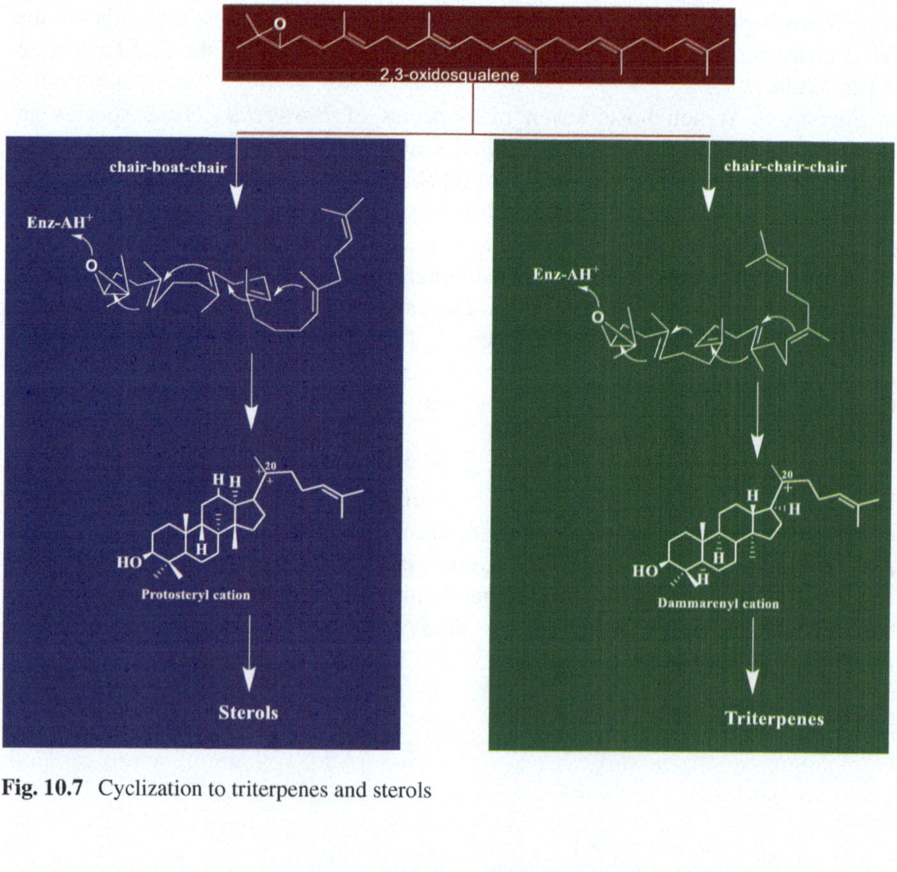

Fig. 10.7 Cyclization to triterpenes and sterols

so-called tailoring enzymes, e.g. cytochrome P450, which furnished tremendous structural diversity. This massive diversity constitutes an infinitesimal percentage of what has been reported from the plant kingdom and an even smaller percentage of what is known from marine and microorganisms.

The cyclization of oxidosqualene is catalysed by oxidosqualene cyclases, which generate either sterol or triterpene scaffolds depending on the conformational arrangement of the 2,3-oxidosqualene (Thimmappa, Geisler, Louveau, O'Maille, & Osbourn, 2014). Interestingly, cyclization commenced at chair-boat-chair conformation results in a protosteryl cation, which in turn furnishes sterols, while the cyclization commenced in a chair-chair-chair conformation results in a dammarenyl cation, which leads to triterpenes (Fig. 10.7) (Thimmappa et al., 2014).

The cyclization reaction catalysed by squalene-hopene cyclase and oxidosqualene cyclases protonates the terminal bond of squalene and epoxy-2,3-oxidosqualene, respectively. Both enzymes are known as class II terpene synthases. The cyclization to mono-, sesqui- and diterpenes is initiated via the ionization of pyrophosphate, which is assisted by Mg2+ cofactors. Similar to other terrestrial plants, the biosynthesis of triterpenes in the *Boswellia* genus commences at (3*S*)-2,3-oxidosqualene. A detailed biosynthetic pathway and the triterpenes isolated from *Boswellia* can be found in Al-Harrasi, Hussain, et al. (2018).

References

Adams, R. (1997). Identification of essential oil components by gas chromatography/mass spectroscopy. *Journal of the American Society for Mass Spectrometry, 6*(8), 671–672.

Al-Harrasi, A., Hussain, H., Csuk, R., & Khan, H. Y. (2018). *Chemistry and bioactivity of boswellic acids and other terpenoids of the genus boswellia*. Amsterdam, The Netherlands: Elsevier.

Al-Harrasi, A., Rehman, N. U., Khan, A. L., Al-Broumi, M., Al-Amri, I., Hussain, J., . . . Csuk, R. (2018). Chemical, molecular and structural studies of Boswellia species: β-Boswellic aldehyde and 3-epi-11β-Dihydroxy BA as precursors in biosynthesis of boswellic acids. PLoS One, 13(6), e0198666.

Ali, N. A. A., Wurster, M., Arnold, N., Teichert, A., Schmidt, J., Lindequist, U., & Wessjohann, L. (2008). Chemical composition and biological activities of essential oils from the oleogum resins of three endemic Soqotraen Boswellia species. *Records of Natural Products, 2*(1), 6.

Basar, S. (2005). *Phytochemical investigations on Boswellia species: Comparative studies on the essential oils, pyrolysates and boswellic acids of Boswellia carterii Birdw., Boswellia serrata Roxb., Boswellia frereana Birdw., Boswellia neglecta S. Moore and Boswellia rivae Engl.* PhD diss., Verlag nicht ermittelbar, University of Hamburg.

Başer, K., Demirci, B., Dekebo, A., & Dagne, E. (2003). Essential oils of some Boswellia spp., myrrh and opopanax. *Flavour and Fragrance Journal, 18*(2), 153–156.

Breitmaier, E. (2006). *Terpenes: flavors, fragrances, pharmaca, pheromones*: John Wiley & Sons, New York.

Camarda, L., Dayton, T., Di Stefano, V., Pitonzo, R., & Schillaci, D. (2007). Chemical composition and antimicrobial activity of some oleogum resin essential oils from Boswellia spp. (Burseraceae). *Annali di Chimica: Journal of Analytical, Environmental and Cultural Heritage Chemistry, 97*(9), 837–844.

Eslamieh, J. (2011). *Cultivation of Boswellia: Sacred Trees of Frankincense; a Book on the History, Cultivation, Original Descriptions, Staging, Hybridization, Holistic Use, and Future of Boswellias in Our World*: A Book's Mind, Fort Collins, CO, USA.

Hamm, S., Bleton, J., Connan, J., & Tchapla, A. (2005). A chemical investigation by headspace SPME and GC–MS of volatile and semi-volatile terpenes in various olibanum samples. *Phytochemistry, 66*(12), 1499–1514.

Hill, R. A., & Connolly, J. D. (2012). Triterpenoids. *Natural Product Reports, 29*(7), 780–818.

Kubmarawa, D., Ogunwande, I. A., Okorie, D. A., Olawore, N. O., & Kasali, A. A. (2006). Constituents of the essential oils of Boswellia dalzielii hutch. From Nigeria. *Journal of Essential Oil Research, 18*(2), 119–120.

Mertens, M., Buettner, A., & Kirchhoff, E. (2009). The volatile constituents of frankincense–a review. *Flavour and Fragrance Journal, 24*(6), 279–300.

Mothana, R. A., Hasson, S. S., Schultze, W., Mowitz, A., & Lindequist, U. (2011). Phytochemical composition and in vitro antimicrobial and antioxidant activities of essential oils of three endemic Soqotraen Boswellia species. *Food Chemistry, 126*(3), 1149–1154.

Murthy, K. S. R., Reddy, M. C., Rani, S. S., & Pullaiah, T. (2016). Bioactive principles and biological properties of essential oils of Burseraceae: A review. *Journal of Pharmacognosy and Phytochemistry, 5*(2), 247.

Niebler, J., & Buettner, A. (2016). Frankincense revisited, part I: Comparative analysis of volatiles in commercially relevant Boswellia species. *Chemistry & Biodiversity, 13*(5), 613–629.

Niebler, J., Eslamieh, J., & Buettner, A. (2016). Frankincense Revisited, Part II: Volatiles in Rare Boswellia Species and Hybrids. *Chemistry & Biodiversity, 13*(5), 630–643.

Paul, M. (2012). Chemotaxonomic investigations on resins of the frankincense species *Boswellia papyrifera*, Boswellia serrata and *Boswellia sacra*, respectively, Boswellia carterii: A qualitative and quantitative approach by chromatographic and spectroscopic methodology. NT - Chemie, Universitat Des Saarlandes, Saarbrücken, Germany.

Provan, G. J., Gray, A. I., & Waterman, P. G. (1987). Monoterpene-rich resins from some kenyan burseraceae. *Flavour and Fragrance Journal, 2*(3), 115–118.

Singh, B., Kumar, R., Bhandari, S., Pathania, S., & Lal, B. (2007). Volatile constituents of natural *Boswellia serrata* oleo-gum-resin and commercial samples. *Flavour and Fragrance Journal, 22*(2), 145–147.

Small, E. (2017). 55. Frankincense and Myrrh–imperilled divine symbols of religion's duty to conserve biodiversity. *Biodiversity, 18*(4), 219–234.

Strappaghetti, G., Corsano, S., Craveiro, A., & Proietti, G. (1982). Constituents of essential oil of *Boswellia frereana*. *Phytochemistry, 21*(8), 2114–2115.

Thimmappa, R., Geisler, K., Louveau, T., O'Maille, P., & Osbourn, A. (2014). Triterpene biosynthesis in plants. *Annual Review of Plant Biology, 65, 225*.

Thulin, M., & Warfa, A. M. (1987). The frankincense trees (Boswellia spp., Burseraceae) of northern Somalia and southern Arabia. *Kew Bulletin, 42*, 487–500.

Tolera, M., Menger, D., Sass-Klaassen, U., Sterck, F. J., Copini, P., & Bongers, F. (2012). Resin secretory structures of Boswellia papyrifera and implications for frankincense yield. *Annals of Botany, 111*(1), 61–68.

Worwood, V. A. (2016). The complete book of essential oils and aromatherapy, revised and expanded: Over 800 natural, nontoxic, and fragrant recipes to create health, beauty, and safe home and work environments: New World Library, Navato, CA, USA.

Conclusion

Frankincense has been traded and used since ancient times, especially in the Arabian Peninsula and African regions. The uses of frankincense range from cultural to religious and from medicinal to economical purposes. However, there are still many questions and myths to address regarding its uses by ancient civilizations and the true impact of its trade on economies and relationships. To this end, using approaches from palaeobotany and economic botany could reveal important information regarding resin samples and their composition and utilization. In addition, the historical perspectives on frankincense use and trade are scattered and should be collected properly and translated.

Frankincense trees (family Burseraceae, order Sapindales; angiosperms) are classified into 19 species from the *Boswellia* genus, and they grow in the Arab and African regions and on the Indian sub-continent. The most important are *B. sacra*, *B. serrata*, *B. papyrifera* and *B. frereana*. During the write-up of this chapter, only a few authentic sources were found for documentation in this book. This result suggests that a great deal of work is needed to understand the taxonomic basis of the *Boswellia* genus, whereas more literature needs to be published online about its classical taxonomy, distribution, habitat, aetiology, ethnobotany, ecology, threats and conservation efforts. Currently, there is more focus on a few species; however, other species in this genus are frequently overlooked and are not very well presented in the literature. In addition, more efforts towards clarification and understanding must be made to differentiate and discriminate among the species of *Boswellia*.

The frankincense produced by *Boswellia* trees has been regarded as the defence-related response of the tree, and this production is an energy-driven process from the primary and secondary metabolism. Frankincense is an essential part of the resin acids, as synthesized via active or passive transport systems into the axial or radial resin canals of the tree and produced constitutively (stored) or de novo during wounding. Frankincense collected by many communities in African countries is marketed as raw products to European and North American countries. The current literature is deficient in uncovering the exact amount of oleo-resin produced in the world, and the equal distribution of benefits to grassroots level and community

A. Al-Harrasi et al., *Biology of Genus Boswellia*,
https://doi.org/10.1007/978-3-030-16725-7

awareness about the harvest are challenging issues in the sustainable supply of frankincense. In addition, the cross-sectional physiology among different species of the genus *Boswellia* and the variations in the chemical profiles of the resin have not yet been fully elucidated. There is a need not only to understand these issues but also to barcode a wider variety of resins across different countries. Furthermore, there are various threats and conservation issues that are being confronted by different populations of species in the genus *Boswellia*, which must be addressed.

The physiology and biosynthesis of resin from *Boswellia* trees has not yet been fully elucidated. Even so, the *B. sacra* and *B. papyrifera* populations have been studied to some extent. These studies provide basic information and explain the response factor elicitation during wounding stress, but the underlying mechanisms have not yet been fully explained. When viewing the current trends in global warming and climatic changes, various populations of the *Boswellia* genus are prone to abiotic stresses. These external pressures along with anthropogenic influences can threaten the existing resources for use by future generations. These studies should support further molecular investigation coupled with conservation efforts. Furthermore, in-depth studies are essential to understanding tree growth, ecophysiology and transcript regulation under a variety of environmental stresses such as heat, drought and high mineral-containing soil. This exploration should not exclude the roles and perturbations of biotic stresses in damaging tree stands.

Regarding biotic organisms, the isolation, characterization and profiling of microbial communities associated with *Boswellia* have been the least reported topics. Trees growing under extreme environmental conditions can act as hubs of important microorganisms that can be used to understand tree ecology and symbiosis with other living organisms but will also help to find the best, most suitable and applicable endemic microbial symbionts for the improved growth and propagation of *Boswellia* trees. These microbes can be endophytic or epiphytic in origin and can be used as a source of plant growth-promoting microbes under both abiotic and biotic stress conditions.

In view of abiotic and biotic stresses, there is a dire need to improve the propagation and tissue culture technology development for *Boswellia* species. Various *Boswellia* species grow under different environmental conditions, such as humid tropical and dry tropical regions. Due to the environmental conditions, the growth conditions will change for their respective species. Similarly, the amount of water and proper nutrients required for *Boswellia* plant growth varies with these environmental conditions. Moreover, tissue culture has the potential to provide a higher multiplication of uniform genotypes in a short time compared to the vegetative practices used in the past. Similarly, *Boswellia* stem and root cuttings have been very successful compared to *Boswellia* seedlings grown in nurseries, which have had poor success rates. In vitro techniques used for the conservation of *Boswellia* species have already been reported by various researchers. However, studies on using alternate culture vessels for the micropropagation of *Boswellia* could be performed. Similarly, *Boswellia* micropropagation is a preliminary effort, and further studies on optimum methods for the higher yield of elite plant clones are the future prospects in this field.

For propagation and conservation purposes, understanding the genomics and the related architecture is essential. At present, only the chloroplast genome of *B. sacra* has been reported. The complete chloroplast genome sequence of *B. sacra* has revealed a circular genome of 160,543 bp. The elucidation of the chloroplast genome could provide additional phylogenetic insights at the taxonomic level, and these results could be used to develop molecular markers. The study of the *B. sacra* cp genome is vital to envisaging fresh, in-depth insights into various aspects of its chloroplast genome evolution and for the further exploration of the genetic diversity of *B. sacra* and the other 18 species in this genus. In addition, whole genome sequencing and the transcriptomics of some of the keystone species would unveil the major mechanisms behind frankincense production, its composition and cueing agents.

The chemical composition of frankincense shows diverse terpene contents in the resins harvested from the 19 species in the *Boswellia* genus. This diversity has greatly attracted the interest of biologists, chemists, pharmacologists and medicinal chemists due to their fascinating structural diversity, interesting structures and exceptional biological activities. There are some interesting variations in the terpene concentrations of the different *Boswellia* species. The essential oils from all the reported species have revealed more than 300 volatile constituents. Moreover, more than 100 diterpenes and triterpenes have been reported from this genus. This result clearly demonstrates the enormous structural diversity of the *Boswellia* genus.

Index

A. Al-Harrasi et al., *Biology of Genus Boswellia*,
https://doi.org/10.1007/978-3-030-16725-7

The manufacturer's authorised representative in the EU is Springer Nature Customer Service Centre GmbH, Europaplatz 3, 69115 Heidelberg, Germany. If you have any concerns regarding our products, please contact ProductSafety@springernature.com

Printed and bound by CPI Group (UK) Ltd, Croydon, CR0 4YY
22/09/2026
02221069-0001